RAND NATIONAL SECURITY RESEARCH DIVISION

T0146379

The Blended Retirement System

Retention Effects and Continuation Pay
Cost Estimates for the Armed Services

Beth J. Asch, Michael G. Mattock, James Hosek

Prepared for the Office of the Secretary of Defense and the U.S. Coast Guard

For more information on this publication, visit www.rand.org/t/RR1887

Library of Congress Control Number: 2017951399
ISBN: 978-0-8330-9791-0

Published by the RAND Corporation, Santa Monica, Calif.
© Copyright 2017 RAND Corporation
RAND® is a registered trademark.

Cover design by Tanya Maiboroda. Images courtesy of Fotolia.

Support RAND
Make a tax-deductible charitable contribution at
www.rand.org/giving/contribute

www.rand.org

Preface

This report presents findings on the effect of the Blended Retirement System (BRS) on active component military retention and reserve component participation. The BRS, created by the National Defense Authorization Act (NDAA) of 2016 and amended by the NDAA of 2017, represents the first major change to the armed and uniformed services' retirement system since the end of World War II. The BRS retains a defined-benefit plan from the legacy system and adds a defined-contribution plan and a new pay called continuation pay (CP). The report includes findings on CP rates and cost; presents BRS retention and cost findings for all armed services; describes our U.S. Coast Guard analysis in detail;[1] and shows the effects of the BRS in the steady state and in the transition to the steady state for all five armed services.

The U.S. Department of Defense requested that RAND use its dynamic retention model (DRM) for enlisted and officer personnel in the Air Force, Army, Marine Corps, and Navy to simulate the retention and cost effects of the BRS. This part of the project was conducted within the RAND National Defense Research Institute's Forces and Resources Policy Center.

The Coast Guard asked RAND to develop a database that permitted estimation of DRM models for enlisted and officer Coast Guard personnel and to use the estimated model to simulate the retention effects and CP cost of the BRS. This part of the project was conducted within the RAND Homeland Security and Defense Center.

The report should interest those concerned with the retention and cost effects of the BRS in general and specifically its effects on each service.

About the RAND National Defense Research Institute

The RAND National Defense Research Institute is a federally funded research and development center, housed within the RAND National Security Research Division, and sponsored by the Office of the Secretary of Defense, the Joint Staff, the Unified Combatant Commands, the Navy, the Marine Corps, the defense agencies, and the defense Intelligence Community. For more information on the Forces and Resources Policy Center, visit www.rand.org/nsrd/ndri/centers/frp or contact the director (the contact information is provided on the web page).

---

[1] In earlier work, RAND provided substantial analysis to the Department of Defense and the Military Compensation and Retirement Modernization Commission to support deliberations that led to the BRS. The prior modeling and results, documented in Asch, Hosek, and Mattock (2014) and Asch, Mattock, and Hosek (2015), did not include the Coast Guard because of data limitations.

About the RAND Homeland Security and Defense Center

The RAND Homeland Security and Defense Center conducts analysis to prepare and protect communities and critical infrastructure from natural disasters and terrorism. Center projects examine a wide range of risk-management problems, including coastal and border security, emergency preparedness and response, defense support to civil authorities, transportation security, domestic intelligence, and technology acquisition. Center clients include the U.S. Department of Homeland Security, the U.S. Department of Defense, the U.S. Department of Justice, and other organizations charged with security and disaster preparedness, response, and recovery. For more information about the Homeland Security and Defense Center, visit www.rand.org/hsdc or contact the director at hsdc@rand.org.

Contents

Figures and Tables

Figures

Tables

Summary

The Blended Retirement System (BRS) replaces the legacy retirement system with a three-part system that includes a defined-benefit (DB) plan, similar in structure to the legacy system; a defined-contribution (DC) plan that vests personnel much earlier than with the DB plan; and an increase in current compensation in the form of continuation pay (CP). CP is paid in the midcareer and is computed as a multiplier times the service member's basic pay at that time. The National Defense Authorization Act (NDAA) of 2016 set a minimum floor for the CP multiplier of 2.5 for active component (AC) personnel and 0.5 for reserve component (RC) personnel, and the services have discretion to increase CP above the floors. Under the BRS, members who are serving as of December 31, 2017, are grandfathered under the legacy system. However, those members with fewer than 12 completed years of service (YOS) (or reservists with fewer than 4,320 points) at the start of BRS implementation on January 1, 2018, will be given the opportunity to opt in to the BRS. Finally, under the BRS, members reaching retirement have the option to receive a part of their DB annuity as a lump sum payment payable immediately upon retirement from the military. Members with 12 or more YOS must stay with the legacy system.

Our analysis supported the deliberations leading to BRS legislation. The analysis was based on the Dynamic Retention Model (DRM) and modules written for simulation, graphics, and costing. This modeling capability provided retention and cost estimates for the Air Force, Army, Marine Corps, and Navy (Asch, Hosek, and Mattock [2014]; Asch, Mattock, and Hosek, [2015]). The DRM is a model of an individual's retention decisions over their active and reserve careers. The DRM accounts for expected military and external earnings, allows for individual differences in their taste for military service and for random shocks in each period, and permits the individual to reoptimize depending on the conditions realized in a period. Models were estimated for each service, separately for officers and enlisted personnel using longitudinal retention data. The estimated model is used to simulate the retention and cost effects of changes to the compensation system, such as the change to the BRS. While the U.S. Department of Defense (DoD) was also interested in results for the U.S. Coast Guard, such results were not feasible given the short time horizon for the analysis and the fact that the data used to estimate the DRM for the four DoD services did not include Coast Guard personnel. Consequently, we were not able to provide analysis of the retention and cost effects of alternative retirement reform proposals for Coast Guard personnel.

This report contains research findings from two related projects, one conducted for DoD and one for the Coast Guard. At the request of DoD, we used our DRM capability to simulate the steady-state effects of the BRS on AC retention,[1] RC participation among those with prior

---

[1] By *steady state*, we mean when all members have spent an entire career under the reformed system.

AC service, and CP costs for both officers and enlisted for the Air Force, Army, Marine Corps, and Navy. We had done analysis for DoD of proposals leading up to the BRS legislation, documented in the reports cited earlier. The new analysis for DoD summarized in this report focuses on estimates of the retention and cost effects of the BRS. At the request of the Coast Guard, we developed a new database tracking the individual careers of Coast Guard personnel, used these data to estimate enlisted and officer DRMs for the Coast Guard, and used the estimated models to simulate the retention and cost effects of the BRS for Coast Guard personnel. In addition, we considered the retention and costs effects of the BRS for AC personnel in all five armed services in the transition to the steady state and provided an estimate of opt-in behavior among enlisted and officer personnel as part of both projects. Thus, this report provides a consolidated set of retention and cost results for the five armed services: Air Force, Army, Coast Guard, Navy, and Marine Corps.

With respect to the DRMs for the Coast Guard, we found that the estimated models predict actual retention behavior for AC personnel. Specifically, we used the Coast Guard DRM estimates to predict the cumulative retention profile by YOS for AC enlisted personnel and officers and compared the predictions with the observed retention profiles in the data. We found that the predicted values fit observed retention profiles well. Our model fits for RC personnel are adequate.

Our results show that the BRS can support a steady-state force and experience mix for all five armed services—Air Force, Army, Coast Guard, Marine Corps, and Navy—that are quite close to the current forces for enlisted personnel and officers in each service. While there is no presumption that future requirements will call for the same size and mix as the baseline force under the legacy retirement system, we assessed the retention effects of the BRS in terms of how well it could achieve the baseline. Also, in our policy simulations for the BRS, we computed the AC and RC CP multipliers that produced the closest fit to baseline retention, given the other parameters of the BRS. We found that these CP multipliers are similar across the services but differ between enlisted personnel and officers. For enlisted personnel, we found that the baseline force is achievable when the CP multiplier is set at or near the floors of 2.5 for AC personnel and 0.5 for RC personnel. For officers, the floor-level CP multipliers do not maintain baseline retention for any service, and higher CP multipliers, close to about one year of basic pay for AC personnel, are required. The results on CP multipliers are information the services can take into account in choosing where to set their multipliers.

In addition to retention effects, the DRM also provides estimates of the changes in costs. Because the DoD actuary and the Coast Guard actuary provide estimates of the change in DB and Thrift Savings Plan (TSP) costs, the focus of our cost analysis was on CP costs. Both DoD and the Coast Guard asked us to provide estimates of CP costs in the steady state and in the transition years.

Table S.1 shows estimated steady-state AC CP costs in millions of 2016 dollars. Steady-state CP costs for AC Coast Guard personnel are $24.6 million when CP multipliers are set at levels to sustain retention for both enlisted personnel and officers. When CP multipliers are set at the floors, AC CP costs are about half, or $12.3 million, although, more importantly, AC officer retention is not maintained. In the case of the four services in DoD, AC CP costs are $293.7 million in total when the multipliers are set at the floor for both officers and enlisted personnel, but $718.1 million when set at the levels that sustain retention in each service for enlisted personnel and officers.

Table S.1
Steady-State Active Component Continuation Pay Costs (in 2016 $M)

	Minimum CP Multiplier	Optimized CP Multiplier
Air Force	85.0	203.1
Army	120.3	265.1
Coast Guard	12.3	24.6
Marine Corps	25.0	60.9
Navy	63.4	189.0

In addition to the CP costs, each armed service will also be required to make TSP contributions on behalf of members, another source of steady-state cost. Offsetting these costs are the savings to the government associated with lower DB payouts.

In the transition to the BRS, we predict that virtually all enlisted personnel in each of the services grandfathered under the legacy system but with three or fewer YOS will opt in to the BRS. These junior members have a low probability of reaching 20 YOS and vesting under the legacy system, so the BRS with earlier vesting in the TSP is more attractive. Beyond YOS three, the percentage of personnel at each YOS that would opt in varies across the services. Those with more YOS have missed out on TSP contributions that would have been made by the service on their behalf had they entered the BRS with fewer YOS. Consequently, all else equal, the BRS is less attractive to those with more YOS. On the other hand, those with fewer than but near 12 YOS are also closer to receiving CP, making the BRS more attractive to those with more YOS.

In the case of enlisted personnel in the Air Force, Army, and Navy, the attraction of CP offsets the lack of TSP contributions in the early career years, and opt-in rates are high until YOS eight or nine, but fall off for those close to YOS 12. However, in the Marine Corps and Coast Guard, the attraction of CP is insufficient to offset the lack of TSP contributions in the early years and the opt-in rate falls among those with YOS closer to 12. Results differ across the services because of differences in retention (and the likelihood of reaching 20 YOS and qualifying for the legacy system), in the optimal CP multiplier that sustains retention, and in the estimated parameters for each service, especially the estimated personal discount rate.

For officers, predicted opt-in rates are relatively high, even for those with more YOS, when the CP multipliers are set high enough to sustain officer retention. As mentioned, we found that the CP multipliers required to sustain officer retention under the BRS, relative to baseline retention, were substantially higher than enlisted CP multipliers. Although the BRS with optimized CP supports the baseline retention profile, an incumbent member is not obliged to opt in and will only do so if the perceived gain is positive. This will be the case for members who are less sure of staying until 20 YOS and qualifying for legacy retirement benefits and who may be attracted by the CP payment at 12 YOS.[2] For many officers, the relatively

---

[2]   NDAA 2017 changed CP to allow the services to pay it at any YOS from eight through 12. Under NDAA 2016, CP required a four-year service obligation, but NDAA 2017 reduced it to a minimum obligation of three years. We analyze the BRS as legislated under NDAA 2016 and do not consider the NDAA 2017 changes. We describe here the features of CP as written in NDAA 2016.

high CP multiplier required to sustain retention is a strong draw to the BRS among those with more YOS, enough to offset the fact that those with more YOS have missed out on TSP contributions that would have been associated with their early careers. Only those quite close to 12 YOS have a low likelihood of opt-in, although the pattern differs somewhat across the services. We found that few officers would opt in to the BRS when CP multipliers for officers are set at minimum levels.

Given the high opt-in rates among officers for those with fewer YOS, but low opt-in rates for officers very near 12 YOS, CP costs in the transition rise quickly in the initial phase of the transition period as those who opt in eventually reach 12 YOS. For example, for the Army, CP costs for AC enlisted and officer personnel start at about $80.4 million in 2018, increase to nearly $217 million in 2019, and gradually increase to $265 million thereafter. In the case of the Coast Guard, CP costs start at about $5 million and increase to nearly $19 million after the first year, and gradually increase to nearly $25 million thereafter.

We also explored the steady-state retention and cost effects of setting a common CP multiplier for enlisted personnel and officers. A common CP multiplier might be desirable to promote equity between officers and enlisted personnel in the elements of the new retirement system. A common multiplier that is above the level required to sustain enlisted retention would result in an increase in the enlisted AC force size in the steady state, an effect that could be mitigated with more stringent application of up-or-out rules. A common multiplier below the level to sustain AC officer retention would reduce officer force size, and additional resources, such as higher basic pay or higher special and incentive pays, would be required to sustain officer retention relative to the baseline line. Consequently, costs increase when the CP multiplier is set to a common level for officers and enlisted personnel relative to when it is set to the optimized levels that sustain officer and enlisted retention separately.

Finally, the DRM capability can be used to assess the retention and cost effects of additional legislative changes to the BRS and aspects of its implementation, such as the retention effects of the BRS for members who make a given lump-sum choice. The capability for the Air Force, Army, Marine Corps, and Navy has also been used in the past to assess the retention and cost effects of compensation changes other than to the military retirement system. The development of a new DRM capability for the Coast Guard offers the opportunity to apply the capability to Coast Guard compensation policy questions in the future.

Acknowledgments

We would like to thank several individuals in the Coast Guard who gave generously of their time to assist us in Coast Guard personnel policies and retention patterns. Specifically, we thank CDR Jeremy Anderson, Chase Grafton, LT Kathryn Walter, LT Matthew Zinn, LCDR Stephan Donley, LT Margaret Ward, and LCDR Roger Robitaille. We are grateful to Coast Guard actuary Richard Virgile for information related to retirement costing. Our Coast Guard action officer provided tremendous support throughout the course of this research, and we especially thank LCDR Corey Braddock and his predecessor, LCDR Matthew Rooney. Scott Seggerman at the Defense Manpower Data Center assisted us in accessing and developing the database for estimating Coast Guard DRMs. At RAND, Arthur Bullock did a terrific job developing the database needed to estimate the Coast Guard DRMs. Finally, we thank Kate Anania at RAND for her help as a research assistant.

For our research on the Air Force, Army, Marine Corps, and Navy, we appreciate guidance received from Jeri Busch, Director of Military Compensation in the Office of the Under Secretary of Defense for Personnel and Readiness, and Vee Penrod, Chief of Staff to the Under Secretary of Defense for Personnel and Readiness. We benefited from the input of Gary McGee, Steve Galing, Patricia Mulcahy, and Don Svendsen within the Directorate of Military Compensation. We are grateful to Joel Sitrin, chief DoD actuary, and Peter Rossi and Peter Zouras of the DoD Office of the Actuary for their help in providing cost estimates of the reforms.

We also thank the reviewers of this report: Matt Baird of RAND; and Curt Gilroy, the director of the 9th Quadrennial Review of Military Compensation and the former director of the Office of Accession Policy within the Office of the Under Secretary of Defense for Personnel and Readiness.

Abbreviations

AC	active component
ACOL	Annualized Cost of Leaving
BAH	basic allowance for housing
BAS	basic allowance for subsistence
BFGS	Broyden-Fletcher-Goldfarb-Shanno
BRS	Blended Retirement System
CP	continuation pay
DB	defined benefit
DC	defined contribution
DHS	U.S. Department of Homeland Security
DMDC	Defense Manpower Data Center
DoD	U.S. Department of Defense
DRM	Dynamic Retention Model
FY	fiscal year
MCRMC	Military Compensation and Retirement Modernization Commission
NDAA	National Defense Authorization Act
OSD	Office of the Secretary of Defense
QRMC	Quadrennial Review of Military Compensation
RC	reserve component
RC/T	reserve component/transit
RMC	regular military compensation

TSP	Thrift Savings Plan
WEX	Work Experience File
YOS	years of service

## CHAPTER ONE

Introduction

The National Defense Authorization Act (NDAA) for fiscal year (FY) 2016, as amended by the NDAA of 2017, made substantial changes to the retirement plan for the armed and uniformed services, including the U.S. Coast Guard. For decades, the services had operated under a defined-benefit (DB) system that vests members after 20 years of service (YOS) in an immediate annuity computed based on years of service and basic pay using a 2.5-percent multiplier. The NDAA created a new retirement system, which became known as the Blended Retirement System (BRS), and continues to include a DB plan but adds two new components: a defined-contribution plan (DC), known as the Thrift Savings Plan (TSP), and continuation pay (CP). The TSP would provide an automatic agency contribution on behalf of service members with additional matching contributions. Because it would vest after two years of service, it would give a retirement benefit to members much earlier than the 20 years required under the legacy system. CP is a retention incentive paid to midcareer members who commit to a service obligation. As a trade-off to adding the TSP and CP components, the NDAA reduced the DB multiplier from 2.5 percent to 2.0 percent. A key role of CP is to provide a retention incentive among those in their midcareers to offset the reduction in retention incentives for midcareer personnel that would accompany the reduced DB multiplier. Members who qualify for the DB have the option to receive part of the DB annuity between their retirement age and age 67 (or the Social Security retirement age) in the form of a lump sum. All new accessions after January 1, 2018, will be automatically enrolled into the BRS. Current serving members are grandfathered into the legacy system, while those with fewer than 12 YOS will have the opportunity to opt in to the BRS.

Motivation for Reform

The BRS represents the first major change in the military retirement system since the end of World War II, although criticisms of the system and suggestions for reform date back almost as far. Various commissions, working groups, reviews, and studies analyzed the system and recommended alternatives, generally citing three major deficiencies.[1] First, the legacy system is considered inequitable because only a minority of military members qualify for retirement benefits, and its vesting at 20 years may be perceived as out of step with civilian employers and the prevalence of 401(k) plans and other portable retirement benefits—a disparity, if not an

---

[1] The list of commissions, reviews, and studies of the retirement system is extensive. Reviews of these studies are provided in Asch, Hosek, and Mattock (2014); Henning (2011); Hudson (2007); Christian (2006); and Warner (2006).

outright inequity. Second, it is viewed as inefficient because it places too much compensation in the form of deferred payments, despite the fact that the typical service member is young and has a preference for current versus deferred compensation. As a result, compensation costs are higher than necessary. Third, it is considered inflexible because the immediate vesting point at 20 YOS induces similar career lengths in all occupational specialties. However, optimal career length may well differ by occupational specialty in light of training costs, the value of on-the-job experience, and the value of specific knowledge about plans, equipment, tactics, policies, and regulations. Yet the legacy system limits such flexibility. In addition, the BRS added to flexibility as viewed by the member, because the defined contribution benefit is portable. The legacy system, with its DB, was inherently not portable.

These reviews also found advantages of the legacy system. It is viewed as having a stabilizing effect on the retention of midcareer personnel, who bring considerable training, experience, and leadership and comprise the pool of candidates for top leadership positions. For service members completing 20 YOS, the legacy system provides funds for a successful transition from the military to a civilian career and additional income during the "second career" in the civilian market.

Recent reviews of the system focused on identifying alternatives that maintained the advantages of the legacy system while addressing the criticisms. In general, the reviews found that a blended plan, sometimes called a *hybrid approach*, achieved these objectives. The U.S. Department of Defense (DoD) working group on military compensation reform, convened from September 2011 to June 2013, recommended that the current military retirement system be modernized with a blended system, and it offered alternatives in which the details of the blended approach varied across the alternatives (DoD [2014]). The details focused on when different elements of the plan would vest, how to reduce the DB plan, how current compensation should be increased, the parameters of the DC plan, and others. The alternatives were forwarded to the Military Compensation and Retirement Modernization Commission (MCRMC), an independent commission mandated by the NDAA for FY 2013. MCRMC also recommended a blended system, and, indeed, many features included in the legislated BRS came from MCRMC recommendations.

Analytic Support and Purpose of the Study

To support its assessment of alternative retirement proposals, the Office of the Secretary of Defense (OSD) asked us for general analytical support, as well as modeling and cost analyses. We have a substantial body of research and analysis related to military compensation and retirement policy, including research in support of numerous previous Quadrennial Reviews of Military Compensation (QRMCs). Past research includes comparisons of military and civilian pay and the development, estimation, and application of a stochastic dynamic programming model, known as the Dynamic Retention Model (DRM) of active and reserve retention. The application of the DRM involves simulations of the impact of compensation and retirement policy changes on active component (AC) and reserve component (RC) retention, as well as on cost and outlays, in the steady state and during the transition to the steady state. We used the DRM approach to simulate the cost and retention effects of military retirement reform alternatives proposed by the DoD working group (Asch, Hosek, and Mattock [2014]) and the

MCRMC (Asch, Mattock, and Hosek [2015]). The DRM of AC retention and RC participation is summarized in Chapter Three, with additional details in the appendix.

OSD requested that we use the DRM to assess the retention and cost effects of the BRS for the Air Force, Army, Marine Corps, and Navy. While we had provided analysis of past proposals leading up to the BRS legislation, documented in the reports cited earlier, DoD lacked quantitative estimates of how the BRS would affect officer and enlisted retention and lacked information on the CP multipliers—and the CP cost—required to sustain retention. The Coast Guard was excluded from the OSD request, however, because of a lack of DRM estimates for this service. We did not estimate Coast Guard models when we first estimated models for the other services because the Defense Manpower Data Center (DMDC) data used to estimate the models for the other services, known as the Work Experience File (WEX), did not include Coast Guard personnel. The Coast Guard asked us to develop a database from the active-duty and reserve-duty master files (the files that form the basis of the WEX) and estimate enlisted and officer DRMs for the Coast Guard, similar to those previously estimated for the Air Force, Marine Corps, and Navy.

The purpose of the research summarized in this report was to use our DRM capability to simulate the steady-state effects of the BRS on AC retention, on RC participation among those with prior AC service, and on CP costs for both officers and enlisted for the Air Force, Army, Marine Corps, and Navy, and to develop a similar DRM capability for the Coast Guard and use that capability to also simulate the effects of the BRS on Coast Guard retention and costs. In addition, the research provided estimates of CP costs in the transition to the steady state. In short, the research provided a set of estimates of the effects of the BRS for the five armed services: Air Force, Army, Coast Guard, Marine Corps, and Navy.

For the analysis of Air Force, Army, Marine Corps, and Navy retention and cost under the BRS, we used the DRM capability developed for the 11th QRMC (DoD [2012]) and documented in past reports (cited earlier in this chapter). For the Coast Guard, we constructed a database for Coast Guard personnel using DMDC data that permitted estimation of DRMs for the Coast Guard. We estimated parameters for enlisted and officer DRMs with these data using contextual background information and input gathered from the Coast Guard about officer and enlisted careers. We then simulated steady-state and transitional retention and cost effects of the BRS for the Coast Guard. For all five services, we provide information on the level of CP that is predicted to sustain enlisted and officer retention relative to a baseline, and provide information on the percentage of grandfathered members predicted to switch to the BRS.

Organization of the Report

Chapter Two describes the features of the BRS. Chapter Three gives a brief overview of the DRM, both the estimation and simulation capability, with more model details provided in the appendix. Chapter Four contains steady-state simulation results, while Chapter Five describes results for the transition period. Chapter Six provides concluding thoughts. The appendix gives details on the data development and model adaptation for the Coast Guard and presents estimates of the model's parameters and information on how well the model fits the observed data.

## CHAPTER TWO

Elements of the Blended Retirement System

The BRS has three main components: a DC plan, a DB plan, and CP. In this chapter, we describe each element and briefly discuss the opt-in feature of the BRS. The main elements of the BRS are summarized in Table 2.1, drawn from DoD (2016).

Defined-Contribution Plan

The DC plan is known as the TSP. Under the BRS, all members joining the armed and uniformed services after January 1, 2018, will be automatically enrolled in the TSP with an automatic member contribution of 3 percent of basic pay. The service will contribute on behalf of the service member, and the service contributions consist of two parts: an automatic contribution of 1 percent of basic pay that begins after 60 days of the start of service, and a matching contribution beginning at the start of the third YOS. The service will match up to 4 percent of basic pay, according to the schedule shown in Table 2.2. Members must contribute 5 percent of basic pay to receive the maximum 4-percent match. Service members opting in to the BRS

Table 2.1
Comparison of the Legacy and Blended Retirement Systems

Plan Element	Legacy	BRS
Defined-benefit vesting	20 years	20 years
Defined-benefit multiplier	2.5%	2.0%
Defined-benefit payment working years		Full annuity or lump-sum option (50% or 25%); RC lump sum based on annuity from age 60 to retirement age
Defined-contribution agency contribution rate		1% automatic; plus up to 4% matching (max = 5%)
Defined-contribution rate YOS		1%: entry + 60 days until 26 YOS Matching: start of 3 YOS–26 YOS
Defined-contribution member contribution rate		3% automatic; full match requires 5% contribution
Defined-contribution vesting		Start of YOS 3
Continuation pay multiplier (months of basic pay)		Minimum 2.5 for AC, 0.5 RC; with additional amount varying
Continuation pay YOS/additional obligation		At 12 YOS with 4-year additional obligation
Opt-in		Must be serving on 1/1/2018 and have less than 12 YOS as of 12/31/17; opt-in period is 1/1/18–12/31/2018

RAND *RR1887-2.1*

Table 2.2
TSP Individual and Agency Automatic and Matching Contribution Rates

Individual Contribution (%)	Agency Automatic Contribution (%)	Agency Matching Contribution (%)	Total TSP Contribution (%)
0	1	0	1
1	1	1	3
2	1	2	5
3	1	3	7
4	1	3.5	8.5
5	1	4	10

SOURCE: Office of the Secretary of Defense (2016).

in 2018 receive both the automatic and matching contributions immediately. Like other DC plans, members have full access to their TSP funds at age 59½.

Continuation Pay

CP is a new element of compensation under the BRS that increases current compensation. The purpose of CP is to sustain the size and experience mix of the force by providing a retention incentive to those in their midcareers to offset the reduction in retention incentives caused by the reduced DB multiplier. The TSP also offsets the reduced DB multiplier, but its effect on midcareer retention is muted, as it is not payable until age 59½, while CP is an increase in compensation at YOS 12. Although the future force may not be the same as the present force, the services and OSD deemed it important that any alternative under consideration be able to achieve the size and mix of the present force, the "baseline" under the legacy retirement system. CP would be in addition to any special and incentive pays currently offered to service members. Members would receive CP at YOS 12. Insofar as members are forward-looking, CP provides an inducement for those with fewer than 12 YOS to stay until year 12. Once members reach YOS 12, they can receive CP provided they make a four-year service obligation. By YOS 16, the incentive to stay provided by the availability of the DB annuity at YOS 20 has become quite strong and few members leave.

CP is a multiple of monthly basic pay. AC members are guaranteed a minimum of 2.5 months of basic pay (i.e., a 2.5 multiplier), and RC members are guaranteed a minimum of 0.5 months of basic pay (i.e., a 0.5 multiplier). A service may pay CP above these minimums, and the service would have to request funds to cover the cost of doing so. The CP multiplier may vary by service and by officer and enlisted personnel for the AC and RC.

CP is not intended to replace existing special and incentive pays that target compensation to service members in recognition of differences in their working conditions, risk of danger, nature of work, specialized skills, and unusual external civilian opportunities. These pays are expected to continue as they have in the past. A number of these pays are intended to help sustain retention in specific occupations, such as medical-related specialties and aviation-related specialties. As the legacy retirement system and basic pay table do not vary with occupation, it is not necessarily the case that CP would vary by occupation, and we do not model CP as varying across personnel within the enlisted or officer force for a given service. That said, the services do have the discretion to allow CP multipliers that are above the minimum to vary across occupational areas.

CP did not vary by occupation in our analysis. As discussed in Chapter Three, CP multipliers were determined during policy simulations as the value producing the best fit to the baseline active and reserve force size and retention profile (cumulative retention probability by years of service), given the other elements of the reform. The multipliers are reported in Chapter Four.

CP entails a four-year service obligation. Members who leave the force before completing their four-year obligation are required to repay CP on a prorated basis. For example, a member who served only one year out of the four would be required to repay three-quarters of CP received at YOS 12.

Revised Defined-Benefit Plan

The revised DB plan has a multiplier of 2.0 percent, down from 2.5 percent under the legacy system. That is, the value of the retirement annuity changes from 2.5 percent × YOS × the average of the highest three years of basic pay, to 2.0 percent × YOS × the average of the highest three years of basic pay. Vesting for the DB plan continues to be upon completion of YOS 20. The legacy system is called the "high-three" system.[1]

Upon AC retirement, members will be offered the option to receive the regular full annuity (2-percent multiplier) immediately or one of two lump-sum payment options—the member may choose either 25 percent or 50 percent of the discounted present value of future retirement benefits up to age 67—along with an offsetting reduced annuity up to age 67 and the regular full annuity thereafter. That is, all individuals would receive an annuity based on the 2-percent multiplier after 67, but, for the period between the age of retirement and age 67, the individual can choose at retirement to take a full annuity (no lump sum) or a reduced annuity with a lump-sum payment.

The legislation creating the BRS directed the Secretary of Defense to consider studies of personal discount rates in setting the discount rate to be used for computing the lump-sum amount. At the time of our study, no announcement had been made about the choice of discount rate to be used. Because of the uncertainty about the discount rate, as well as our inability to model to lump-sum choice—a limitation of this implementation of the DRM that we discuss in Chapter Three—our analysis assumes that all members choose the full annuity and do not choose either of the lump-sum options.

---

[1] Technically, there are three legacy systems in place as a result of modifications to the system in 1981 and 1988. Pre-1981 entrants receive a fully inflation-protected annuity that is computed based on final basic pay. Those who entered between 1981 and 1986 are under the "high-three" system, in which the retirement annuity is fully inflation-protected but based on the individual's high three years of basic pay rather than final basic pay. The Military Retirement Reform Act of 1986, known as REDUX, changed the annuity formula to

$$(0.40 + 0.035 \times YOS - 20) \times \text{high-three average pay for the years between separation and age 62.}$$

At age 62, retired pay reverted to the high-three formula. REDUX also changed the inflation protection. As part of NDAA 2000, members at YOS 15 who were covered by REDUX were given a choice to stay under REDUX and receive a $30,000 bonus or be covered by the high-three system.

Opt In

A final feature of the BRS concerns the transition to the new plan. All currently serving members and retirees are grandfathered under the existing system, but those with 12 or fewer YOS (or those reservists with fewer than 4,320 points) have the choice to opt in to the new system between January 1, 2018, and December 31, 2018. All new members who enter after January 1, 2018, will be automatically enrolled in the new system. In Chapter Three, we simulate the transition to the steady state for the AC, including the opt-in choice. We assume AC members choose to opt in if the value of staying in the AC at the time of the choice is higher under the new system than under the existing system. In our simulation computer code, only those with 12 or fewer YOS are permitted the opt-in choice.

CHAPTER THREE

Brief Overview of the Dynamic Retention Model, Its Limitations, and Its Optimization Approach

Our DRM is well suited to the analysis of structural changes in military compensation, such as the changes under the BRS. The model's capability has steadily increased; for example, new, faster estimation and simulation programs have been written, costing has been refined, and the model can now show retention and cost effects in both the steady state and the year-by-year transition to the steady state. The approach is documented in several RAND reports (e.g., Mattock, Hosek, and Asch [2012], which was prepared for the 11th QRMC, and Asch, Hosek, and Mattock [2013]). We provide a more detailed description of the DRM, as well as more details of our application of the DRM to the Coast Guard, including data used, model estimates, and model fits, in the appendix. In this chapter, we provide a brief overview.

The DRM is based on a mathematical model of individual decisionmaking over the life cycle of the individual in a world with uncertainty and in which members have heterogeneous preferences (tastes) for active and reserve service. Model parameters are estimated using data on military careers drawn from administrative data files. The model begins with service in the AC, and individuals make a stay/leave decision in each year. Those who leave the AC take a civilian job and, at the same time, choose whether to participate in the RC. The decision of whether to participate in the RC is made each year, and the individual can move into or out of the RC from year to year. More specifically, a reservist can choose to remain in the RC or leave the RC to lead a purely civilian life, and a civilian can choose to enter the RC or remain a civilian. In the model, each service has a single RC.

A key parameter in the model is the personal discount factor. The discount factor indicates how much a member values a dollar today versus one year in the future. For the Air Force, Army, Marine Corps, and Navy, we have estimated separate models for each service, for both officers and enlisted service members. One of the estimated parameters is the personal discount factor. The estimated real personal discount factor ranges from 0.88 to 0.90 for enlisted personnel across the four services. That is, a dollar next year is worth 88 to 90 cents today. For officers, the estimates are similar across services, at 0.94. These estimates, together with the other model estimates, are discussed in past RAND reports, such as in Asch, Hosek, and Mattock (2014, Appendix E).[1] For the Coast Guard, we found that we achieved better results if we assumed a personal discount factor rather than estimating it.[2] We assumed a personal

---

[1] The MCRMC final report cites personal discount rates from past RAND analyses (MCRMC [2015, pp. 33–34]). The rates cited are arithmetic means of the implied rates from the personal discount rates that we estimate.

[2] The Coast Guard model converged with values of the personal discount factor too high to be credible. This might have resulted from changes in force size and shape over our data period, as both the enlisted and officer populations became

discount factor of 0.88 for Coast Guard enlisted personnel and 0.94 for officers. These choices were guided by the estimates found for the other services.

The data for estimating the DRM for the Air Force, Army, Marine Corps, and Navy were from the DMDC WEX. The WEX contains person-specific longitudinal records of AC and RC service members that track individual service member careers in the AC and RC from entry. We used the WEX data of service members who began their military service in 1990 or 1991 and tracked their individual careers in the AC and, if they join, the RC through 2010, providing 21 years of data on 1990 entrants and 20 years on 1991 entrants. For enlisted personnel and for officers in each service, we drew samples of 25,000 individuals who entered the AC in FYs 1990 and 1991, constructed each service member's history of AC and RC participation, and used these records in estimating the model.

We supplemented these data with information on active, reserve, and civilian pay. AC pay, RC pay, and civilian pay are averages based on the individual's years of AC, RC, and total experience, respectively. We used 2007 military pay tables. Military pay increases are typically across-the-board, with the structure of pay by grade and year of service remaining the same.[3] Therefore, we did not expect our results to be sensitive to the choice of year. Data on regular military compensation (RMC) and basic pay were from the *Selected Military Compensation Tables*, also known as the Greenbook (Office of the Under Secretary of Defense for Personnel and Readiness, Directorate of Compensation [2007]). For civilian pay opportunities for enlisted personnel, we used the 2007 median wage for full-time male workers with associate's degrees. For officers, we used the 2007 80th-percentile wage for full-time male master's degree holders in management occupations. All data on civilian pay opportunities are from the U.S. Census Bureau (DeNavas-Walt, Proctor, and Smith [2008]).

The estimation methodology and model estimates for the Air Force, Army, Marine Corps, and Navy are reported in the documents cited at the start of this chapter. Also, in the case of the Air Force, we separately estimated models for Air Force rated and nonrated officers because the retention profiles for these two groups are different and the rated community is a major subset of the Air Force officer community.[4] Consequently, we show results for enlisted personnel and officers for the Air Force, Army, Marine Corps, and Navy, but, in the case of the Air Force, we show two sets of results for officers, for the nonrated and rated communities.

As mentioned in Chapter One, the WEX does not include Coast Guard personnel. Consequently, we used quarterly DMDC active-duty and reserve-duty master files from 1990 to 2015 and created longitudinal data files for Coast Guard enlisted personnel and officers comparable to the WEX data. The data files we created track Coast Guard members who began their military service in 1990 through 2007 and track their individual careers in the AC and, if they join, the RC, through 2015, providing 25 years of data on 1990 entrants. We used these

---

larger and more senior. The increase in seniority, which probably resulted from personnel management decisions, could have manifested in the estimates as a higher personal discount factor, indicating that personnel were more patient than estimated for the other services. Constraining the discount factor to the median estimated for the other services resulted in models that fit the data well.

[3] An exception was the structural adjustment to the basic pay table in FY 2000 that gave larger increases to midcareer personnel who had reached their pay grades relatively quickly (after fewer YOS). A second exception was the expansion of the basic allowance for housing (BAH), which increased in real value from FY 2000 to FY 2005. It should be noted that the costing analysis is in 2016 dollars.

[4] Rated officers include pilots, combat systems officers, air battle managers, and remotely piloted aircraft officers.

constructed service histories of Coast Guard personnel to estimate enlisted and officer DRMs for Coast Guard personnel. As mentioned, the estimation methodology and model estimates for the Coast Guard are discussed in more detail in the appendix.

We also developed simulation code that allowed us to simulate retention over the military career in the AC and RC and to compute the cumulative retention profile in the steady state. We simulated the retention profile under the current compensation system, which we call the *baseline force*. We then simulated retention under the BRS. The simulations under the legacy system and under the BRS are scaled to the sizes of the 2009 AC enlisted and officer forces for each service.[5]

Another feature of our simulation capability is the computation of personnel costs. However, in the analysis we present here, we focus on cost of BRS CP, given that DoD and Coast Guard actuaries compute DB and TSP costs.

Simulations were done for the steady state for the AC and RC and for the transition to the steady state for the AC. The transition analysis allowed us to address questions about how the new system will affect members currently in service versus new members who are automatically enrolled in the new system. In our transition analysis, we modeled the choice of existing members to opt in to the new system. We assumed AC members would elect to opt in if the value of staying in the AC at the time of the choice was higher under the new system than under the existing system. This allowed us to compute the percentage of members who would opt in. In our transition modeling, we computed CP costs in the transition to the steady state. We also computed retention during the transition, although, because the BRS with optimal CP was designed to sustain retention, retention was, by and large, the same during the transition to the steady state as it was in the steady state.

Limitations, Advantages, and Assumptions

The DRM has several limitations. The model assumes that real military pay, promotion policy, and real civilian pay do not vary over time, and it excludes demographic factors such as gender, marital status, and spousal employment. It excludes health status and health care benefits, and we do not model deployment or deployment-related pay.

That said, the estimated models fit the observed data reasonably well for the both the AC and the RC. Furthermore, on a more general level, the DRM approach has several rich and realistic features that make it well suited for analyzing the retention and cost effects of the BRS. It is a life-cycle model in which retention decisions are made over an entire career. Those decisions are based on forward-looking behavior that depends on current and future military and civilian compensation. The model allows for uncertainty in future periods on both the military side and the civilian side and recognizes that people may change their minds in the future as they get more information about the military and their external opportunities. It also recognizes that individuals differ in their preferences for service in the actives or in the reserves and accounts for these differences. Furthermore, the model is formulated in terms of the parameters that underlie the retention and reserve participation decision processes rather

---

[5] The year 2009 was chosen at the time when the Air Force, Army, Marine Corps, and Navy DRMs were created in support of the 11th QRMC as a relatively recent year. The choice of year for computing force size is somewhat arbitrary and other years could be chosen instead.

than on the average response of members to a particular compensation policy. Consequently, it is structured to enable assessments of alternative compensation systems that have yet to be tried in both the steady state and the transition to the steady state.

The DRM approach has advantages over alternative retention modeling approaches. Goldberg (2002) provides an extensive discussion of the history of retention models, while Gotz (1990) provides a detailed discussion of the advantages over other approaches. A common alternative is the so-called Annualized Cost of Leaving (ACOL) approach.[6] It is a multiperiod model of retention decisionmaking and could be estimated with regression programs available when it was introduced in the late 1970s, a time when no routines existed to estimate such dynamic programming models as DRM. However, ACOL's tractability comes at a cost. It selects a single future year when it is best to leave the military. Decisionmakers behave as if they know with certainty when they will leave the military, so they are repeatedly surprised by random factors in each future period, although random factors always occur. Said differently, ACOL does not permit the decisionmaker to reoptimize depending on the conditions realized in a future period. From a practical standpoint, the approach can lead to implausible predictions about the retention effects of certain pay policies, thereby leading to flawed policy recommendations, and it cannot be used to predict opt-in behavior to a new policy, as is the case under the BRS. The DRM approach addresses these drawbacks. Apart from ACOL, other retention models were one-period models and were not structured to handle retention behavior over a career or to have forward-looking decisions.

It is also important to recognize some limitations of our modeling that are specific to simulating the BRS reform. The DRM does not model members' choices regarding an annuity or lump sum for AC or RC members. The DRM also does not model members' savings decisions and therefore their decisions regarding whether and how much to contribute to the TSP. Therefore, we are not able to simulate what percentage of members will choose a full annuity versus a partial lump sum or a full lump sum. We are also not able to simulate the distribution of contribution rates among service members to TSP and therefore the average the TSP match rate.[7]

We managed these limitations by assigning all members the same assumed choice in the simulation. In the case of the TSP contribution rate, we assumed members contribute 5 percent of their basic pay, thereby receiving the full 4-percent DoD match rate, on top of the 1-percent automatic contribution.[8] We used the same assumption in our earlier analysis of the BRS for the other services. We also used the same assumption in our earlier analysis with respect to the lump-sum choice. Here, as there, we assumed all members chose the annuity and did not elect a lump sum.

---

[6] In its simplest form, ACOL is the difference in the present value of the income stream to be had from leaving immediately and the income stream from staying s more years in service, put on an annualized basis. It is formulated as the maximum of the expected value of staying and the value of leaving. In contrast, the DRM is formulated as the expected value of the maximum of the value of staying and the value of leaving.

[7] The DRM could be extended to include these decisions. However, data are not available to allow an empirical implementation of this extension.

[8] Prior analysis showed that retention effects were similar under the assumption that all members contributed a lower percent, e.g., 3 percent instead of 5 percent, and with optimized CP multipliers With a 5-percent contribution, the member realizes the greatest gain from the TSP, and the incentive to do this is strong because of dollar-for-dollar matching up to 3 percent and half-dollar-for-dollar matching at 4 percent and 5 percent. With higher TSP contributions from the service, the optimal CP multiplier is slightly lower (Asch, Mattock, and Hosek[2015]).

Having a choice of a lump sum or annuity is a valuable feature of the reform package. Similarly, the availability of a DoD matching contribution is a valuable feature. This additional choice improves the value of staying in the military and therefore improves retention. However, data are not yet available for us to extend the DRM to include these choices and estimate parameters pertaining to them. As a result, we cannot quantify the added value of having the choices or treat this in our simulations. Given that the value of these choices is omitted, we understate the retention effect of the reform package by an unknown amount.[9]

Optimization

As discussed in Chapter Two, CP for the AC and RC equals a CP multiplier times the active-duty monthly basic pay at YOS 12, with a payback feature for those who separate before completing four additional YOS. An important objective of the current analysis is to assess the amount of CP that each service must pay to sustain the baseline retention profiles for enlisted personnel and officers.

To achieve this in our model, our simulations compute optimized values of the continuation pay multiplier, given the other features of the BRS. This involves computing the value of the multiplier that minimizes the difference between the baseline retention profile under the legacy high-three retirement system and the profile under the BRS, given the other features of the BRS, including our assumed TSP contribution rate and annuity choice.

A key question is, what is the relevant baseline? Ideally, the baseline retention profile should reflect the service's required experience mix and force. The baseline we used is the simulated retention profile under the current compensation system and high-three retirement system. While there is no presumption that future requirements will call for the same size and mix as the baseline force, we assessed the retention effects of the BRS in terms of how well it could achieve the baseline.

We show the optimized multipliers in Chapter Four, where we discuss steady-state results.

---

[9] Furthermore, were the value of these choices included, the optimized CP multiplier probably would be somewhat smaller.

Steady-State Retention and Cost Results

This chapter shows our steady-state simulation results. We also show retention results for two cases. The first assumes the CP multiplier is set at the floor of 2.5 for AC personnel and 0.5 for RC. The second uses the optimized values of the CP multiplier—optimized to replicate the baseline retention profile under the BRS. We also present estimates of steady-state CP costs. A key finding is that the optimized enlisted CP multiplier is approximately equal to the 2.5 floor, or slightly higher, depending on the service, while the optimized officer multiplier is substantially higher, implying that a higher CP multiplier is required to sustain baseline officer retention. CP costs are higher when the multipliers are at their optimized values rather than at their floors, but the gain from optimized multipliers is being able to sustain retention at baseline level.

Retention Results Under the Blended Retirement System with Continuation Pay Multipliers Set at Floors

Figures 4.1–4.11 show results by service for enlisted and officer personnel, respectively, assuming the CP multiplier is set at the floors of 2.5 for AC and 0.5 for RC. The charts on the left show AC results. The ones on the right show results for RC personnel with prior AC experience. In all graphics in this chapter, the black line is the retention profile under the legacy (baseline) system, while the red line is the profile under the BRS.

For enlisted personnel, AC retention is close to baseline retention when the CP multipliers for AC and RC are set at the floor. RC participation among those with prior AC service is also close to the baseline. Because CP is one of three elements working simultaneously—the DB, the TSP, and CP—it is more accurate to say that, given the parameters of the DB and TSP plans and of CP (formula, when paid), the minimum floor for CP generally performs well to sustain retention. As we will show later in this chapter, the optimized CP multipliers for enlisted personnel are close to the floor, so it is unsurprising that the floor performs so well.

For officers, when the CP multipliers are set at the floor AC retention falls short of the baseline after around the fifth YOS. In the first five YOS or so, officers are under a service obligation. On the other hand, RC participation among those with prior AC service increases after the fifth YOS under the BRS with minimum CP multipliers. Thus, under the minimum, fewer officers reach 20 YOS in the AC. Lower AC retention in the years approaching 20 YOS indicates that, at the minimum CP multipliers, the BRS offers less gain to staying than the legacy system, and some of those who leave prior to 20 YOS join the RC and qualify for retirement there, as suggested by the pre–20-YOS increase in RC participation. Because of the drop in AC

Figure 4.1
Enlisted Retention Under Legacy (Baseline) System Versus Blended Retirement System at Continuation Pay Multiplier Floors for the Active Component and the Reserve Component: Air Force

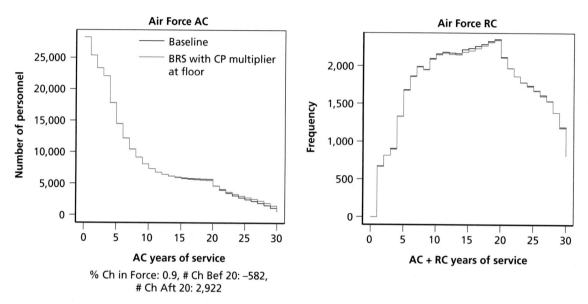

% Ch in Force: 0.9, # Ch Bef 20: –582,
# Ch Aft 20: 2,922

NOTE: "% Ch in Force" is the percentage change in the steady-state force size; "# Ch Bef 20" is the change in the steady-state number of personnel serving with fewer than 20 YOS; "# Ch Aft 20" is the change in the steady-state number of personnel serving with 20 or more YOS. These changes are relative to the 2009 force size, as shown in Table A.3.

RAND RR1887-4.1

Figure 4.2
Enlisted Retention Under Legacy (Baseline) System Versus Blended Retirement System at Continuation Pay Multiplier Floors for the Active Component and the Reserve Component: Army

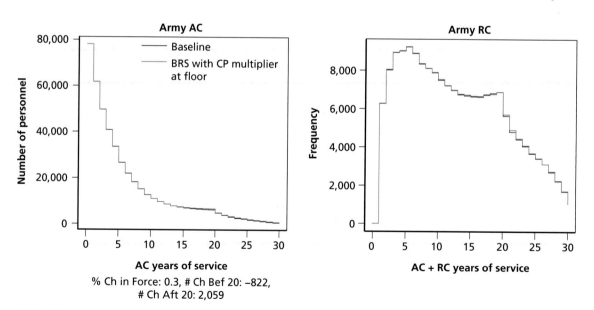

% Ch in Force: 0.3, # Ch Bef 20: –822,
# Ch Aft 20: 2,059

NOTE: "% Ch in Force" is the percentage change in the steady-state force size; "# Ch Bef 20" is the change in the steady-state number of personnel serving with fewer than 20 YOS; "# Ch Aft 20" is the change in the steady-state number of personnel serving with 20 or more YOS. These changes are relative to the 2009 force size, as shown in Table A.3.

RAND RR1887-4.2

Figure 4.3
Enlisted Retention Under Legacy (Baseline) System Versus Blended Retirement System at Continuation Pay Multiplier Floors for the Active Component and the Reserve Component: Coast Guard

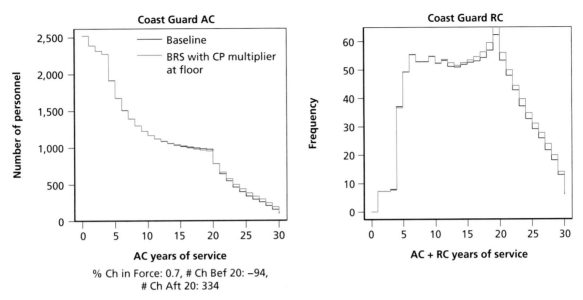

% Ch in Force: 0.7, # Ch Bef 20: –94,
# Ch Aft 20: 334

NOTE: "% Ch in Force" is the percentage change in the steady-state force size; "# Ch Bef 20" is the change in the steady-state number of personnel serving with fewer than 20 YOS; "# Ch Aft 20" is the change in the steady-state number of personnel serving with 20 or more YOS. These changes are relative to the 2009 force size, as shown in Table A.3.

RAND RR1887-4.3

Figure 4.4
Enlisted Retention Under Legacy (Baseline) System Versus Blended Retirement System at Continuation Pay Multiplier Floors for the Active Component and the Reserve Component: Marine Corps

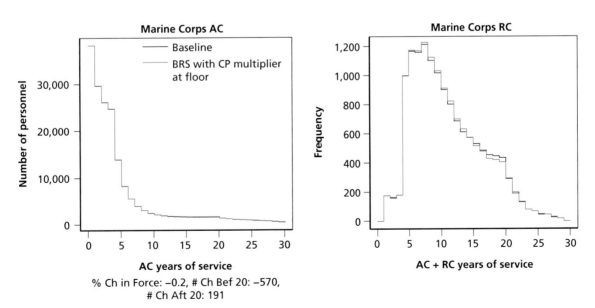

% Ch in Force: –0.2, # Ch Bef 20: –570,
# Ch Aft 20: 191

NOTE: "% Ch in Force" is the percentage change in the steady-state force size; "# Ch Bef 20" is the change in the steady-state number of personnel serving with fewer than 20 YOS; "# Ch Aft 20" is the change in the steady-state number of personnel serving with 20 or more YOS. These changes are relative to the 2009 force size, as shown in Table A.3.

RAND RR1887-4.4

Figure 4.5
Enlisted Retention Under Legacy (Baseline) System Versus Blended Retirement System at Continuation Pay Multiplier Floors for the Active Component and the Reserve Component: Navy

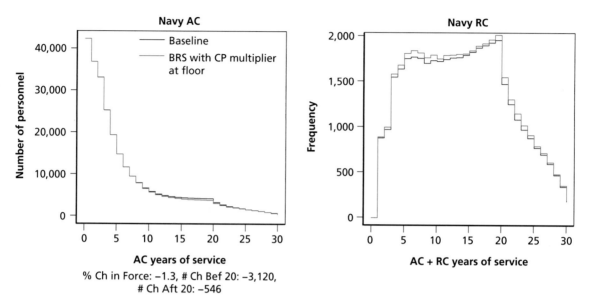

% Ch in Force: –1.3, # Ch Bef 20: –3,120,
# Ch Aft 20: –546

NOTE: "% Ch in Force" is the percentage change in the steady-state force size; "# Ch Bef 20" is the change in the steady-state number of personnel serving with fewer than 20 YOS; "# Ch Aft 20" is the change in the steady-state number of personnel serving with 20 or more YOS. These changes are relative to the 2009 force size, as shown in Table A.3.

RAND *RR1887-4.5*

Figure 4.6
Officer Retention Under Legacy (Baseline) System Versus Blended Retirement System at Continuation Pay Multiplier Floors for the Active Component and the Reserve Component: Air Force Nonrated

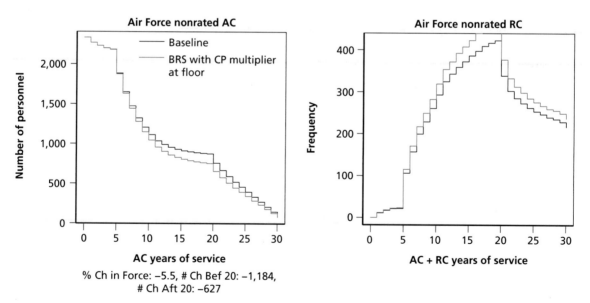

% Ch in Force: –5.5, # Ch Bef 20: –1,184,
# Ch Aft 20: –627

NOTE: "% Ch in Force" is the percentage change in the steady-state force size; "# Ch Bef 20" is the change in the steady-state number of personnel serving with fewer than 20 YOS; "# Ch Aft 20" is the change in the steady-state number of personnel serving with 20 or more YOS. These changes are relative to the 2009 force size, as shown in Table A.3.

RAND *RR1887-4.6*

Figure 4.7
Officer Retention Under Legacy (Baseline) System Versus Blended Retirement System at Continuation Pay Multiplier Floors for the Active Component and the Reserve Component: Air Force Rated

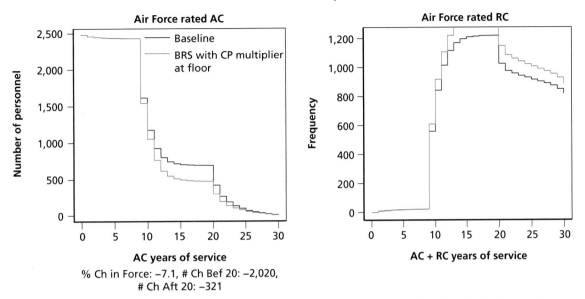

% Ch in Force: −7.1, # Ch Bef 20: −2,020,
# Ch Aft 20: −321

NOTE: "% Ch in Force" is the percentage change in the steady-state force size; "# Ch Bef 20" is the change in the steady-state number of personnel serving with fewer than 20 YOS; "# Ch Aft 20" is the change in the steady-state number of personnel serving with 20 or more YOS. These changes are relative to the 2009 force size, as shown in Table A.3.

RAND RR1887-4.7

Figure 4.8
Officer Retention Under Legacy (Baseline) System Versus Blended Retirement System at Continuation Pay Multiplier Floors for the Active Component and the Reserve Component: Army

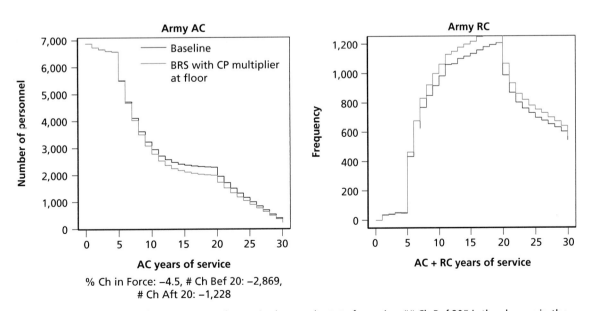

% Ch in Force: −4.5, # Ch Bef 20: −2,869,
# Ch Aft 20: −1,228

NOTE: "% Ch in Force" is the percentage change in the steady state force size; "# Ch Bef 20" is the change in the steady-state number of personnel serving with fewer than 20 YOS; "# Ch Aft 20" is the change in the steady-state number of personnel serving with 20 or more YOS. These changes are relative to the 2009 force size, as shown in Table A.3.

RAND RR1887-4.8

Figure 4.9
Officer Retention Under Legacy (Baseline) System Versus Blended Retirement System at Continuation Pay Multiplier Floors for the Active Component and the Reserve Component: Coast Guard

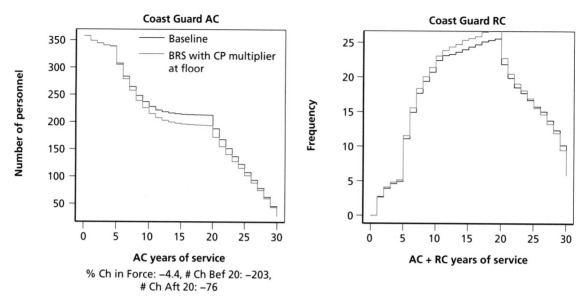

% Ch in Force: –4.4, # Ch Bef 20: –203,
# Ch Aft 20: –76

NOTE: "% Ch in Force" is the percentage change in the steady state force size; "# Ch Bef 20" is the change in the steady-state number of personnel serving with fewer than 20 YOS; "# Ch Aft 20" is the change in the steady-state number of personnel serving with 20 or more YOS. These changes are relative to the 2009 force size, as shown in Table A.3.

RAND RR1887-4.9

Figure 4.10
Officer Retention Under Legacy (Baseline) System Versus Blended Retirement System at Continuation Pay Multiplier Floors for the Active Component and the Reserve Component: Marine Corps

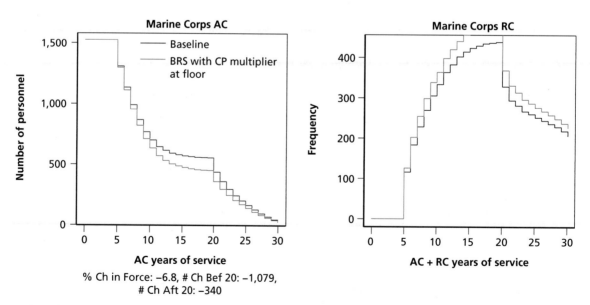

% Ch in Force: –6.8, # Ch Bef 20: –1,079,
# Ch Aft 20: –340

NOTE: "% Ch in Force" is the percentage change in the steady-state force size; "# Ch Bef 20" is the change in the steady-state number of personnel serving with fewer than 20 YOS; "# Ch Aft 20" is the change in the steady-state number of personnel serving with 20 or more YOS. These changes are relative to the 2009 force size, as shown in Table A.3.

RAND RR1887-4.10

Figure 4.11
Officer Retention Under Legacy (Baseline) System Versus Blended Retirement System at Continuation Pay Multiplier Floors for the Active Component and the Reserve Component: Navy

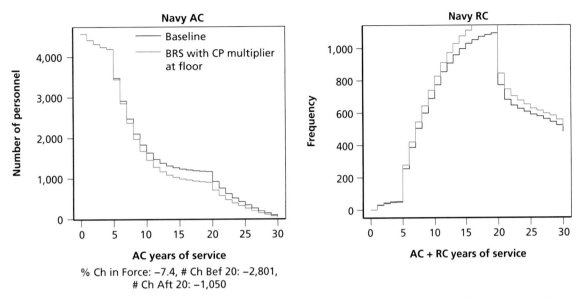

% Ch in Force: –7.4, # Ch Bef 20: –2,801,
# Ch Aft 20: –1,050

NOTE: "% Ch in Force" is the percentage change in the steady-state force size; "# Ch Bef 20" is the change in the steady-state number of personnel serving with fewer than 20 YOS; "# Ch Aft 20" is the change in the steady-state number of personnel serving with 20 or more YOS. These changes are relative to the 2009 force size, as shown in Table A.3.

RAND RR1887-4.11

retention prior to YOS 20, fewer personnel also would serve after YOS 20 in the AC. Overall, we find that steady-state AC force size decreases for the officer force in each service when CP is set at the minimum level. The reduction varies across services, from 4.4 percent for Coast Guard officers to 7.4 percent for Navy officers.

Retention Results Under the Blended Retirement System with Optimized Continuation Pay Multipliers

Table 4.1 shows CP multipliers that minimize the difference between the baseline retention profile under the legacy retirement system and the profile under the BRS for enlisted personnel and officers in each service. We compute these optimized CP multipliers with the DRM simulation capability. The figures in bold are optimized values below the 2.5 and 0.5 floors for AC and RC multipliers, respectively, set by the NDAA 2016 legislation. Figures 4.1–4.11 correspond to Figures 4.12–4.22, showing the AC retention and RC participation results for enlisted personnel and officers, respectively, under the BRS when we optimized the CP multipliers.

For both enlisted and officer personnel, AC retention and RC participation are close to baseline retention when the CP multipliers are optimized. As shown in Table 4.1, the optimized values for enlisted personnel are close to the legislative floors. Consequently, as is the case when the CP multiplier is set to the floor, AC enlisted retention under the BRS with optimized CP multipliers replicates the profile under the legacy retirement system, as shown in Figures 4.1–4.5. For officers, the results are much different under the BRS with optimized

Table 4.1
Active Component and Reserve Component Optimized Continuation Pay Multipliers for Enlisted Personnel and Officers, by Service

	AC	RC
Enlisted		
Air Force	**2.05**	0.89
Army	**2.39**	**0.45**
Coast Guard	**2.44**	0.54
Marine Corps	3.40	0.82
Navy	3.97	0.58
Officer		
Nonrated Air Force	13.39	5.99
Rated Air Force	10.27	5.74
Army	10.85	5.62
Coast Guard	11.34	4.11
Marine Corps	9.71	4.62
Navy	12.88	5.94

NOTE: Optimization involves computing the value of the multiplier that minimizes the difference between the baseline retention profile under the legacy high-three retirement system and the retention profile under the BRS, given the other BRS features, including our assumed TSP contribution rate and annuity choice. Figures in bold indicate multipliers that are below the floor set by NDAA 2016.

multipliers. Unlike the cases shown in Figures 4.6–4.11, in which the floor is used, baseline retention and experience mix are now sustained for each service. The AC retention profiles and RC participation profiles under the BRS closely follow the baseline profiles for each service.

Table 4.1 shows that the optimized CP multipliers for officers are far higher than the 2.5 and 0.5 floors for AC and RC, respectively. The implication is that the CP required to sustain officer retention relative to the baseline when the DB multiplier is reduced from 2.5 percent to 2.0 percent must be higher than the floor, despite the presence of the TSP in the BRS.

The table also shows that across the services, optimized CP multipliers for officers are higher than for enlisted personnel, which are close to the floor. For enlisted personnel, the AC multipliers are typically about 2.5 for AC personnel, but vary from 2.05 for the Air Force to 3.97 for the Navy. For officers, the AC multipliers are typically around 11 and vary from 9.21 for the Marine Corps to 13.39 for nonrated officers in the Air Force. The commonality of the optimized CP multipliers within the enlisted and within the officer forces regardless of service is related to the commonality of factors across services within each group. These factors were listed earlier and included differences between enlisted personnel and officers in the parameter estimates, especially the estimated personal discount factors, lower basic pay of enlisted personnel than officers, and lower retention and a lower likelihood of reaching YOS 20 for enlisted personnel. While there are clearly differences across services among enlisted personnel and among officers in all of these factors, the differences are apparently fewer than between officers and enlisted.

Figure 4.12
Enlisted Personnel Under Legacy System Versus Blended Retirement System with Optimized Continuation Pay Multipliers: Air Force

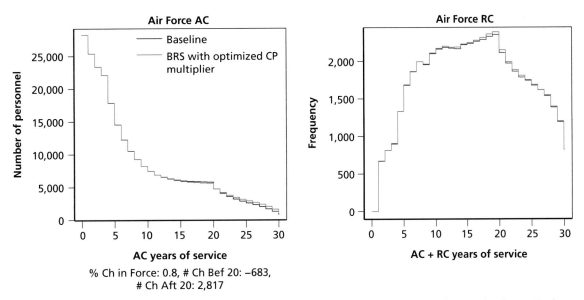

% Ch in Force: 0.8, # Ch Bef 20: −683,
# Ch Aft 20: 2,817

NOTE: "% Ch in Force" is the percentage change in the steady-state force size; "# Ch Bef 20" is the change in the steady-state number of personnel serving with fewer than 20 YOS; "# Ch Aft 20" is the change in the steady-state number of personnel serving with 20 or more YOS. These changes are relative to the 2009 force size, as shown in Table A.3.

RAND RR1887-4.12

Figure 4.13
Enlisted Personnel Under Legacy System Versus Blended Retirement System with Optimized Continuation Pay Multipliers: Army

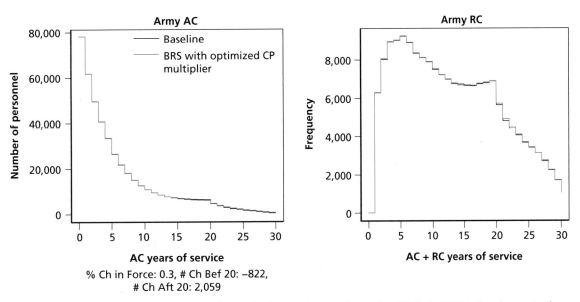

% Ch in Force: 0.3, # Ch Bef 20: −822,
# Ch Aft 20: 2,059

NOTE: "% Ch in Force" is the percentage change in the steady-state force size; "# Ch Bef 20" is the change in the steady-state number of personnel serving with fewer than 20 YOS; "# Ch Aft 20" is the change in the steady-state number of personnel serving with 20 or more YOS. These changes are relative to the 2009 force size, as shown in Table A.3.

RAND RR1887-4.13

Figure 4.14
Enlisted Personnel Under Legacy System Versus Blended Retirement System with Optimized Continuation Pay Multipliers: Coast Guard

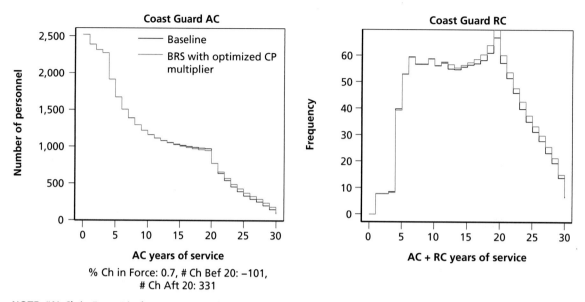

% Ch in Force: 0.7, # Ch Bef 20: −101,
# Ch Aft 20: 331

NOTE: "% Ch in Force" is the percentage change in the steady-state force size; "# Ch Bef 20" is the change in the steady-state number of personnel serving with fewer than 20 YOS; "# Ch Aft 20" is the change in the steady-state number of personnel serving with 20 or more YOS. These changes are relative to the 2009 force size, as shown in Table A.3.

RAND RR1887-4.14

Figure 4.15
Enlisted Personnel Under Legacy System Versus the Blended Retirement System with Optimized Continuation Pay Multipliers: Marine Corps

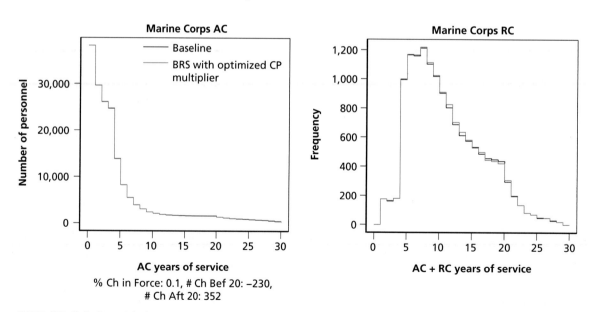

% Ch in Force: 0.1, # Ch Bef 20: −230,
# Ch Aft 20: 352

NOTE: "% Ch in Force" is the percentage change in the steady-state force size; "# Ch Bef 20" is the change in the steady-state number of personnel serving with fewer than 20 YOS; "# Ch Aft 20" is the change in the steady-state number of personnel serving with 20 or more YOS. These changes are relative to the 2009 force size, as shown in Table A.3.

RAND RR1887-4.15

Figure 4.16
Enlisted Personnel Under Legacy System Versus Blended Retirement System with Optimized Continuation Pay Multipliers: Navy

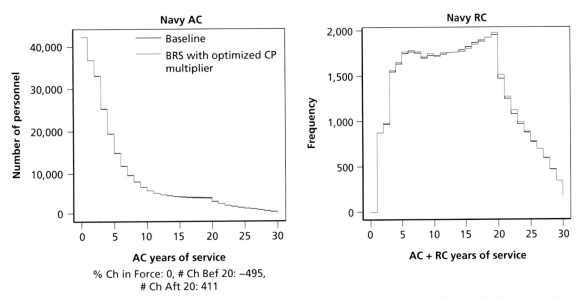

% Ch in Force: 0, # Ch Bef 20: –495,
# Ch Aft 20: 411

NOTE: "% Ch in Force" is the percentage change in the steady-state force size; "# Ch Bef 20" is the change in the steady-state number of personnel serving with fewer than 20 YOS; "# Ch Aft 20" is the change in the steady-state number of personnel serving with 20 or more YOS. These changes are relative to the 2009 force size, as shown in Table A.3.

RAND RR1887-4.16

Figure 4.17
Officers Under Legacy System Versus Blended Retirement System with Optimized Continuation Pay Multipliers: Air Force Nonrated

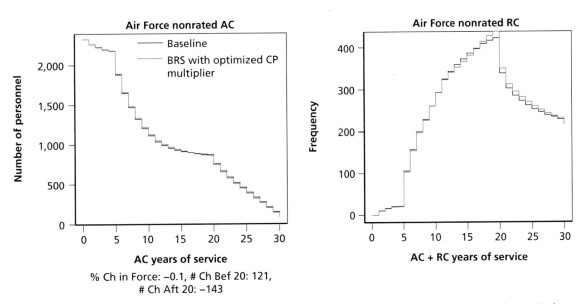

% Ch in Force: –0.1, # Ch Bef 20: 121,
# Ch Aft 20: –143

NOTE: "% Ch in Force" is the percentage change in the steady-state force size; "# Ch Bef 20" is the change in the steady-state number of personnel serving with fewer than 20 YOS; "# Ch Aft 20" is the change in the steady-state number of personnel serving with 20 or more YOS. These changes are relative to the 2009 force size, as shown in Table A.3.

RAND RR1887-4.17

Figure 4.18
Officers Under Legacy System Versus Blended Retirement System with Optimized Continuation Pay Multipliers: Air Force Rated

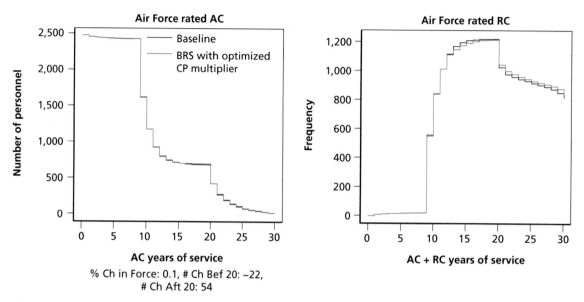

% Ch in Force: 0.1, # Ch Bef 20: –22,
# Ch Aft 20: 54

NOTE: "% Ch in Force" is the percentage change in the steady-state force size; "# Ch Bef 20" is the change in the steady-state number of personnel serving with fewer than 20 YOS; "# Ch Aft 20" is the change in the steady-state number of personnel serving with 20 or more YOS. These changes are relative to the 2009 force size, as shown in Table A.3.

RAND RR1887-4.18

Figure 4.19
Officers Under Legacy System Versus Blended Retirement System with Optimized Continuation Pay Multipliers: Army

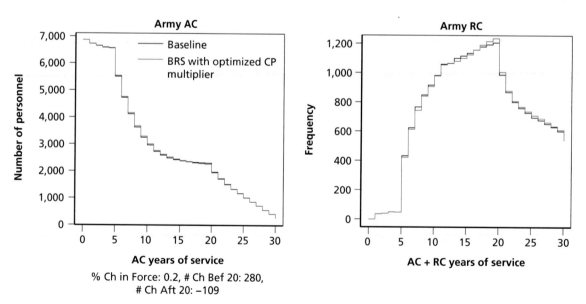

% Ch in Force: 0.2, # Ch Bef 20: 280,
# Ch Aft 20: –109

NOTE: "% Ch in Force" is the percentage change in the steady-state force size; "# Ch Bef 20" is the change in the steady-state number of personnel serving with fewer than 20 YOS; "# Ch Aft 20" is the change in the steady-state number of personnel serving with 20 or more YOS. These changes are relative to the 2009 force size, as shown in Table A.3.

RAND RR1887-4.19

Figure 4.20
Officers Under Legacy System Versus Blended Retirement System with Optimized Continuation Pay Multipliers: Coast Guard

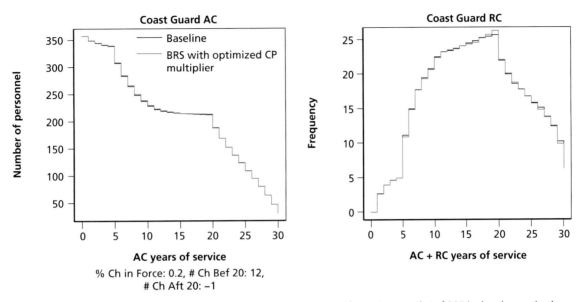

% Ch in Force: 0.2, # Ch Bef 20: 12,
# Ch Aft 20: −1

NOTE: "% Ch in Force" is the percentage change in the steady-state force size; "# Ch Bef 20" is the change in the steady-state number of personnel serving with fewer than 20 YOS; "# Ch Aft 20" is the change in the steady-state number of personnel serving with 20 or more YOS. These changes are relative to the 2009 force size, as shown in Table A.3.

RAND RR1887-4.20

Figure 4.21
Officers Under Legacy System Versus Blended Retirement System with Optimized Continuation Pay Multipliers: Marine Corps

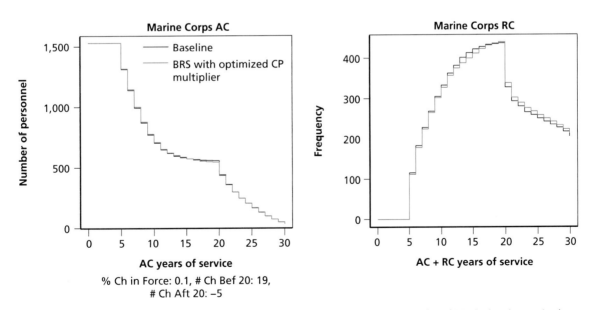

% Ch in Force: 0.1, # Ch Bef 20: 19,
# Ch Aft 20: −5

NOTE: "% Ch in Force" is the percentage change in the steady-state force size; "# Ch Bef 20" is the change in the steady-state number of personnel serving with fewer than 20 YOS; "# Ch Aft 20" is the change in the steady-state number of personnel serving with 20 or more YOS. These changes are relative to the 2009 force size, as shown in Table A.3.

RAND RR1887-4.21

Figure 4.22
Officers Under Legacy System Versus Blended Retirement System with Optimized Continuation Pay Multipliers: Navy

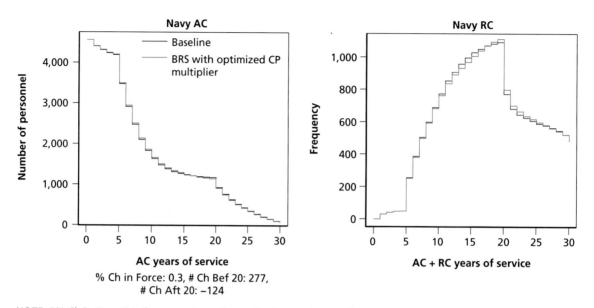

% Ch in Force: 0.3, # Ch Bef 20: 277,
# Ch Aft 20: –124

NOTE: "% Ch in Force" is the percentage change in the steady-state force size; "# Ch Bef 20" is the change in the steady-state number of personnel serving with fewer than 20 YOS; "# Ch Aft 20" is the change in the steady-state number of personnel serving with 20 or more YOS. These changes are relative to the 2009 force size, as shown in Table A.3.

RAND RR1887-4.22

This difference between enlisted personnel and officers in optimized CP multipliers is attributable to several factors. First, the BRS reduces the DB multiplier from 2.5 percent to 2.0 percent. An individual in early career or midcareer and looking forward toward military retirement will expect a lower DB annuity under the BRS. Because officers have higher retention and a higher likelihood of reaching 20 YOS, they are more likely to experience the reduction in the DB multiplier than their enlisted counterparts. Second, officers earn higher basic pay and can expect a larger DB annuity if they become eligible at 20 YOS. Consequently, the reduction in the DB multiplier has a larger effect for officers than for enlisted with a given number of YOS. Finally, we estimate differing parameters for officers and enlisted, and, in particular, different personal discount factors. We typically estimate a personal discount factor for officers of 0.94 and for enlisted of between 0.88 and 0.90.[1] This implies that a dollar one year from now is worth more today for officers—$0.94—than for enlisted personnel—$0.88 to $0.90. Thus, a given reduction in future benefits, such as the reduction from the reduced DB multiplier under the BRS, has more of an effect on today's value for officers than for enlisted personnel. Both CP and TSP help to offset the negative effects of the reduced DB multiplier, but the optimized CP multiplier must be higher for officers than for enlisted personnel.

---

[1]　See Mattock, Hosek, and Asch (2012); Asch, Hosek, and Mattock (2013); and Asch, Hosek, and Mattock (2014). For Coast Guard personnel, the discount factors were assumed based on the estimates of the other services.

Table 4.2
Steady-State Active Component Continuation Pay Costs (in 2016 $M)

	Minimum CP Multiplier	Optimized CP Multiplier[a]
Enlisted		
Air Force	56.5	56.5
Army	79.1	79.1
Coast Guard	8.8	8.8
Marine Corps	15.4	21.3
Navy	41.8	68.2
Officer		
Air Force	28.6	146.7
Army	41.2	186.0
Coast Guard	3.5	15.8
Marine Corps	9.6	39.5
Navy	21.6	120.8
Total cost[b]		
Air Force	85.0	203.1
Army	120.3	265.1
Coast Guard	12.3	24.6
Marine Corps	25.0	60.9
Navy	63.4	189.0

[a] CP costs in the optimized multiplier case are the same as in the minimum CP case for enlisted personnel in the Army and Navy because the optimized multipliers fall below the minimum.

[b] Because of rounding, totals cost do not exactly equal the sum of enlisted and officer costs.

Steady-State Cost

Our simulation capability includes a costing module that enables us to compute BRS-related costs. Because the DoD and Coast Guard actuaries compute DB and DC retirement-related costs, the focus of our costing was on CP costs. Table 4.2 shows our estimates of steady-state AC CP costs for enlisted personnel and officers with optimized CP multipliers and at the floor, the minimum CP multiplier. All dollar figures are in 2016 dollars. By law, the CP multipliers cannot fall below the floors, so Table 4.2 sets the CP multiplier to 2.5 for AC and 0.5 for RC even in the cases in which the optimized values fall below the floor. For example, CP costs for the Coast Guard in the optimized case assume the CP multipliers are 2.5 and 0.5 for AC and RC enlisted personnel, respectively, and not 2.44 and 0.54.

The total cost of CP in the steady state for AC personnel was $12.3 million for the Coast Guard when CP multipliers were set at the floor of 2.5, with nearly three-quarters of the cost attributable to enlisted personnel. But, as shown in Figures 4.6–4.11, officer retention was not sustained relative to the baseline when the CP multiplier was at the floor. At the optimized CP multiplier, officer retention was sustained, as seen in Figures 4.17–4.22, but CP costs were higher; $15.8 million rather than $3.5 million. In total, to sustain retention given the other BRS elements, AC CP costs for the Coast Guard were $24.6 million.

We found similar qualitative results for the Air Force, Army, Marine Corps, and Navy, although the magnitudes differ from that of the Coast Guard. Total cost of CP in the steady state for AC personnel varied from $25 million for the Marine Corps to $120.3 million for the Army across the four services when CP multipliers were set to the minimum. Across the four DoD services, total CP costs in the steady state were $293.7 million in 2016 dollars. But when CP multipliers were optimized to sustain the force, total CP costs were higher, e.g., $60.9 million for the Marine Corps and $265.1 million for the Army. Total CP costs for DoD were $718.1 million in the steady state.

We did not compute the costs of the other elements of the BRS, specifically TSP costs and DB retirement costs, because the DoD and Coast Guard actuaries estimate them. In addition to the CP costs shown in Table 4.2, each armed service will also be required to make TSP contributions on behalf of members, another source of steady-state cost. Offsetting these costs are the savings to the government associated with lower DB payouts. Under the BRS, the DB annuity is computed based on a 2.0-percent multiplier instead of the 2.5-percent multiplier under the legacy system.

In DoD, retirement costs are funded on an accrual basis, so the DB cost of the BRS to the Air Force, Army, Marine Corps, and Navy is the accrual charge, computed as a percentage of the basic pay bill. Because of the lower DB multiplier under the BRS, the steady-state accrual charge is lower, thereby producing a cost savings to each of these services. Coast Guard retired pay is not funded on an accrual basis but is a mandatory pay-as-you-go account, thus the lower DB multiplier does not produce similar cost savings in the near term. Consequently, the Coast Guard will see a cost increase in the near term associated with funding the TSP and CP under the BRS. But it will produce cost savings in the long term as more, and eventually all, members retire under the BRS.

Thus, while the CP cost estimates in Table 4.2 show a source of cost increase associated with the BRS, the lower DB annuity will produce a reduction in costs in the long run. In Chapter Six, we discuss the time pattern of costs in the transition period.

## CHAPTER FIVE

Transition Results

Given military careers of up to 30 years, and in some cases longer, it would take at least 30 years to reach the new steady state as a result of a policy change. Because of concerns related to the near-term effects of a policy change, policymakers are also interested in the effects of the BRS in the transition to the steady state.

The BRS legislation grandfathers existing members in the legacy system, while those entering the military as of January 1, 2018, are covered by the BRS rules and benefits. Grandfathering is often desirable because policymakers do not want to break the implicit contract with existing members and want to ensure that "promises are kept." The NDAA language grandfathers members under the legacy system but permits those with fewer than 12 YOS at the beginning of 2018 to opt in or elect the BRS over the legacy system. The opt-in approach has two potential advantages: Faith is not broken because those who decide to change do so only if they expect to be better off under the new policy, and cost savings can be realized sooner as more members opt in.[1]

We were asked to use an extended version of the DRM developed in past RAND research to predict the percentage of members who will opt in to the BRS and to assess the effects of the BRS on retention and CP costs in the transition period given opt-in behavior. The extended DRM is detailed in Asch, Mattock, and Hosek (2013) and provides results on the effects for the AC only.

To model opt-in behavior, we assume members will opt in if the expected value of staying in the AC is greater under the new system than in the legacy system. When we simulate the transitional effects on retention and costs, we track the retention behavior of each cohort of currently serving members, as well as new entrants automatically placed into the BRS, and do so in calendar time. A member's YOS in the year of BRS implementation, 2018, defines his or her cohort under the BRS, and the number of individuals in that cohort who opt in to the new system depends on the features of the BRS and personal choice, which the simulation handles. The simulation keeps track of each individual's retention experience under the baseline and going forward under the new system, given the individual's opt-in decision. So, the simulation keeps track of individuals in each cohort as they move through their careers, including calendar time, and it aggregates across individuals in different cohorts to produce the cross-sectional force profile for each calendar year. The cross-sectional profiles, together with information on

---

[1] Government outlays are initially higher under the BRS because of service expenditures for CP and the TSP matching contributions, but they will eventually be lower because of the decrease in DB retirement. As the BRS phases in, expenditures initially increase because of CP and the TSP, but later, as more and more service members retire under BRS DBs, expenditures decrease and ultimately produce a significant, persistent decrease in expenditures relative to the legacy system. When more members opt in to the BRS, the point when expenditures decrease relative to baseline comes sooner.

the cost of different elements of the reform, are inputs to our computations of year-by-year cost during the transition. Because the simulation tracks cohorts by calendar year, we are also able to compute the percentage of a cohort that opts in by YOS and track that percentage over time among members of the cohort remaining in the military.

The chapter begins with results on the percentage of currently serving members who opt in to the BRS and then discusses CP cost savings over the transition years.

Percentage of Members Who Opt In

Figure 5.1 shows the percentage of each enlisted cohort in each service estimated to opt in during the 2018 election period, in which *cohort* is defined as the member's YOS in 2018. Thus, "YOS cohort = 10" means the member reached YOS 10 in 2018. Figure 5.2 shows officer opt-in percentages by YOS cohort by service when the CP multiplier for officers is set at the minimum floor of 2.5 for AC and 0.5 for RC. Likewise, Figure 5.3 shows officer opt-in percentages when the CP multiplier for officers is the optimized level to sustain the baseline force size and experience mix, as shown in Table 4.1. In cases in which the optimized CP level falls below 2.5—in the Air Force, Army, and Coast Guard—the CP floor is used instead.

The opt-in rate is 100 percent for those in their first year, by design because new entrants are automatically enrolled into the BRS. Beyond the first YOS, opt-in rates are 100 percent for at least the first few years,[2] but, more generally, they are affected by several factors that can operate to increase or decrease the likelihood of opt-in and the importance of these factors can vary by YOS cohort and between enlisted personnel and officers. Those who opt in at later YOS miss out on the TSP contributions that would have been made on their behalf by the agency had they spent an entire career under the BRS, so opt-in rates will tend to be lower for those facing an opt-in decision later in their careers. On the other hand, we find that higher CP is an inducement to opt in, and it looms larger when individuals are closer to YOS 12 in their careers and face the opt-in decision. Furthermore, the pull of CP is larger for officers because officers have a higher personal discount factor than enlisted personnel so officers do not discount future dollars at as high a rate as enlisted personnel. A final factor affecting the opt-in rate is the likelihood of reaching 20 YOS and being able to reap the benefits of the legacy system—and also how that likelihood changes as members progress through their careers. Those early in their careers have a lower chance of reaching 20 YOS, so the BRS is relatively attractive. Individuals later in their careers have a higher chance of reaching 20 YOS, so opt-in is less attractive, but retention rates tend to stabilize in the midcareer. Once individuals reach about ten YOS, most will reach YOS 20. Thus, the differential effect of retention on the opt-in decision at different YOS is less important for those in their midcareers (around ten to 12 YOS).

For enlisted personnel, opt-in rates are high for those in the initial part of their careers for all of the services. For the Air Force, Army, and Navy, opt-in rates remain high for those with fewer than eight YOS, but rates decline with YOS cohort thereafter. For the Coast Guard and Marine Corps, opt-in rates decline gradually beginning with the fourth YOS cohort for the

---

[2]   Opt-in rates may prove to be less than 100 percent as a result of factors not modeled in the DRM. Such factors might include limited financial literacy, limited understanding of the terms of BRS, and the possibility that a beginning service member does not adapt well to the military and expects to leave before completing the first term of service.

Figure 5.1
Percentage of Enlisted Personnel Who Opt In to Blended Retirement System, by Years-of-Service Cohort

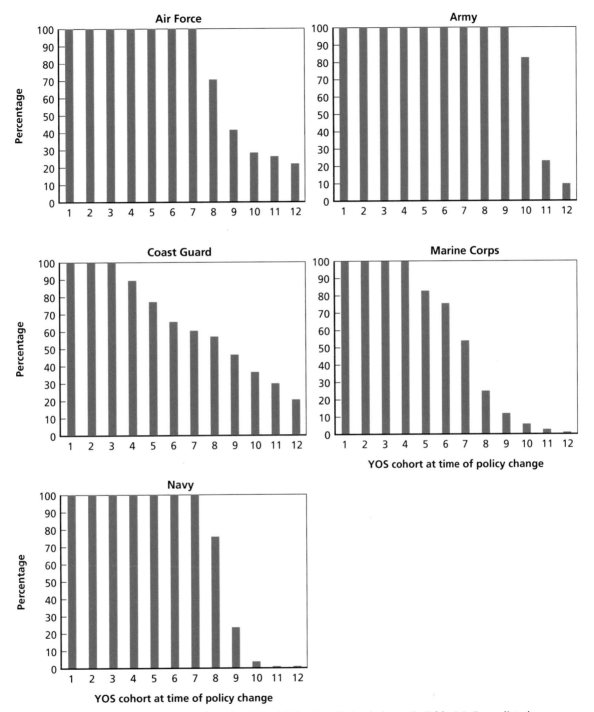

NOTE: The figure shows opt-in rates when the CP multiplier is optimized, shown in Table 4.1. For enlisted personnel, the optimized levels are generally quite close to the minimum floor.

RAND RR1887-5.1

Figure 5.2
Percentage of Officers Who Opt In to Blended Retirement System with Minimum Continuation Pay Multipliers, by Years-of-Service Cohort

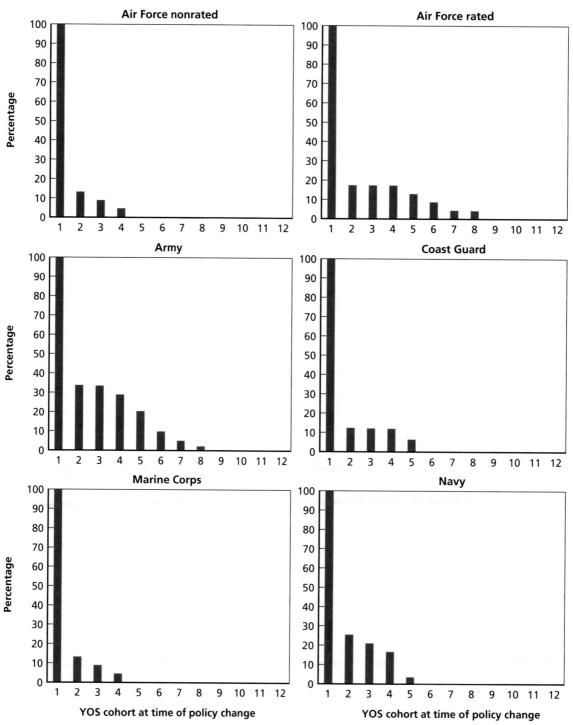

Figure 5.3
Percentage of Officers Who Opt In to Blended Retirement System with Optimized Continuation Pay Multipliers, by Years-of-Service Cohort

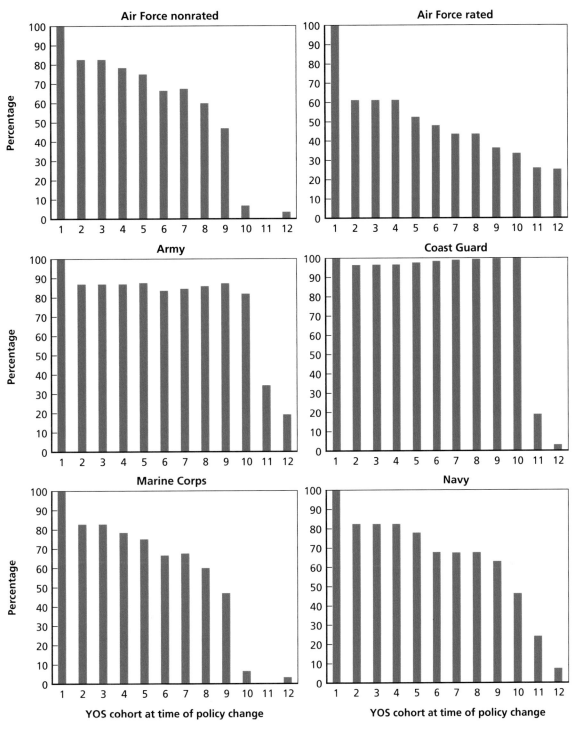

Coast Guard and fifth YOS for the Marine Corps. In general, those who face the opt-in decision later in their careers did not receive TSP contributions in the earlier part of their careers, thereby reducing the value of the BRS relative to the legacy system. The pull of the CP multiplier at YOS 12 offsets this effect. This pull varies across services because it is affected by the cumulative retention rate to YOS 12, the CP multiplier, and the enlisted personal discount rate. Enlisted personnel have relatively low personal discount factors compared to officers, although the estimated rate varies by service. The low discount factor reduces the inducement effect of CP on opt-in for enlisted personnel.

Like enlisted personnel, officers facing the opt-in decision closer to YOS 12 also have missed out on early career TSP contributions. When the CP multiplier is set at the minimum level, the amount of CP is not sufficient in any service to generate high levels of opt-in (Figure 5.2), despite the higher personal discount factor of officers. When the CP multiplier is at the minimum, the opt-in rates are about 10 to 20 percent for those with fewer than six YOS, and are generally zero for those with more than five YOS. The exception is Army officers, but even in this case, opt-in rates are about 30 percent for those with between two and four YOS, dropping off significantly thereafter. On the other hand, CP has a strong inducement effect when it is set at the optimized level that sustains baseline retention for officers. The CP effect is large enough to offset the negative effect of fewer years of TSP contributions among those with more YOS, so opt-in rates are substantially higher than in the minimum CP case (Figure 5.3). Opt-in rates vary by service but are around 80 percent or higher among those with fewer than five YOS and remain high among those with more YOS in the Army and Coast Guard until ten YOS. In the other services, opt-in rates fall gradually through YOS nine or ten. Opt-in rates drop off steeply at around YOS 11 and 12. Officers with 11 or 12 YOS have virtually a 100 percent chance of reaching 20 YOS (Figure 5.2), so the legacy system is quite attractive, despite the inducement effect of CP.

Active Component Retention During the Transition

We can also compute the change in AC retention during the transition period and the composition of personnel under the legacy versus the new system. The retention of personnel under each system and how it changes over time will affect the time pattern of costs. For example, because of the outlays for CP and the TSP, more opt-ins and more individuals under the new system early into the transition will mean that the services will experience a more dramatic near-term increase in costs for these components of the new system.

In this section, we first show the time pattern of overall retention, including personnel under both the legacy system and the BRS, and then the composition of personnel under each system. These results are produced under the assumption that CP is set at the optimized level to sustain retention. We then show the time pattern of overall retention when CP is not set at the optimized level. We show these patterns for officers in two services, one service in DoD— the Air Force nonrated officers—and the Coast Guard. We chose the Air Force nonrated officers because the optimized CP multiplier is highest for this service (13.39), while the opt-in rates fall gradually between YOS two and 12. The optimized CP multiplier is lower for Coast Guard officers (11.34), while opt-in rates are relatively higher between YOS two and 12.

We find that if CP is set to sustain baseline retention, retention is, by and large, constant in the transition to the steady state, as well as in the steady state. This is because the optimized

CP largely succeeds in maintaining force size and experience mix. But if CP is set at the minimum for officers, retention will fall in the transition.

If CP is optimized to sustain retention, it is no surprise that, in conjunction with the other features of the BRS, force size and shape remain virtually the same under the reform. This is the case in the steady state, as shown in Chapter Four, and in the transition to the steady state, as illustrated in Figure 5.4 for nonrated Air Force officers and Coast Guard officers. The figures appear to show only one line—the retention profile for officers—but, in fact, it shows 30 separate lines. Each line represents the retention profile in year *s*, where *s* is the number of

Figure 5.4
Simulated Active Component Nonrated Air Force Officer and Coast Guard Officer Retention During the Transition, Optimized Continuation Pay

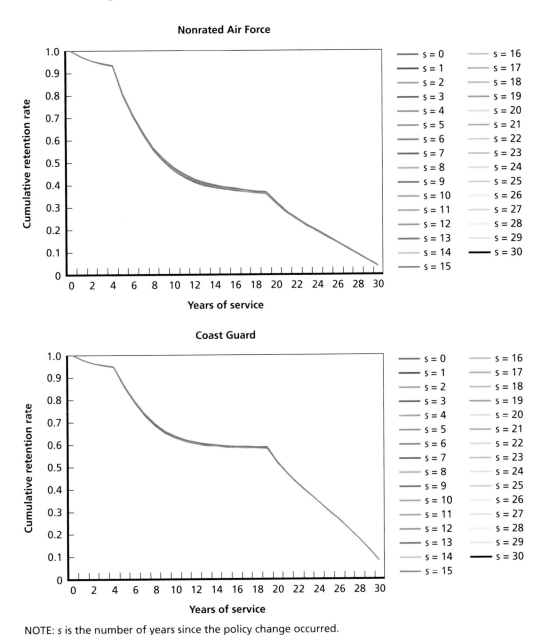

NOTE: *s* is the number of years since the policy change occurred.
RAND RR1887-5.4

years since the policy change occurred, e.g., 2018. Year 0 is the steady state in the baseline, and year 30 is the new steady state. The fact that all of the lines are identical in Figure 5.4 means that retention is unchanged in the transition period.

Although overall retention stays unchanged as time elapses, as shown in Figure 5.5, the composition of personnel does change in terms of who is and is not covered by the new policy, as shown in Figures 5.5 and 5.6. Figure 5.5 shows the retention profiles over the calendar year (or year since reform) of only those in the new system, while Figure 5.6 shows the retention profiles by YOS and years since reform for those who remain under the legacy system. In the baseline ($s = 0$), all members are covered by the legacy system and no one is covered by the new system, hence the flat blue line along the x-axis in Figure 5.5, which is for members who opt in. The counterpart to this figure is Figure 5.6, where the blue line is the steady-state retention curve reflecting that all members are under the legacy system at $s = 0$. In the first year ($s = 1$), new entrants are covered by the new system, while everyone else either remains under the legacy system or opts in to the BRS.

As mentioned earlier, the composition of personnel under the legacy versus the new system and how it changes over time will affect the time pattern of costs. Insofar as the new system incurs higher current costs because of outlays for CP and the TSP, more opt-ins and more individuals under the new system early into the transition will mean that costs will increase more dramatically in the near term.[3] On the other hand, if relatively few individuals opt in, the near-term increase in costs over time will be more gradual. Figure 5.7 shows that officer AC retention falls during the transition if CP is not set at a high enough level to sustain retention. The figure shows the case when CP is set at the minimum of 2.5 for AC personnel. For example, after ten years ($s = 10$), retention is lower among those with fewer than ten YOS because these individuals entered service after the BRS policy change including the minimum CP multiplier and were automatically enrolled in the BRS. Retention among those with more than ten YOS at $s = 10$ reflects both opt-in rates and the subsequent retention of those who opted in versus those who did not opt in. After 30 years since the policy change ($s = 30$), the new steady state is reached with lower retention across the force.

Continuation Pay Costs in the Transition

Using the opt-in information and retention profiles in the transition, we can compute AC CP costs for enlisted personnel and officers in the transition years. Figure 5.8 shows estimates of BRS CP costs for AC personnel for enlisted personnel and officers by service over a ten-year planning period, from 2018 to 2028.

The pattern of costs over the ten-year period and the differences between enlisted personnel and officers reflect the opt-in behavior shown in Figures 5.1 and 5.3. Although the officer force is smaller, CP costs are higher for officers because their basic pay is higher and the optimized CP multiplier is higher. Officer CP costs increase rapidly in all but the Air Force.

For example, for the Army, CP costs increase from about $80.4 million to about $217 million between 2018 and 2019, reflecting the large difference in opt-in rates between YOS 12 and YOS ten in Figure 5.3. About 20 percent of officers in their 12th YOS elect to opt in, so CP costs in the first year are quite low, $62.5 million for officers and $80.4 million for the Army

---

[3]   In the long term, total costs decrease because of lower DB benefits under the BRS.

Figure 5.5
**Simulated Nonrated Air Force and Coast Guard Active Component Retention During the
Transition for Officers Who Opt In, Optimized Continuation Pay**

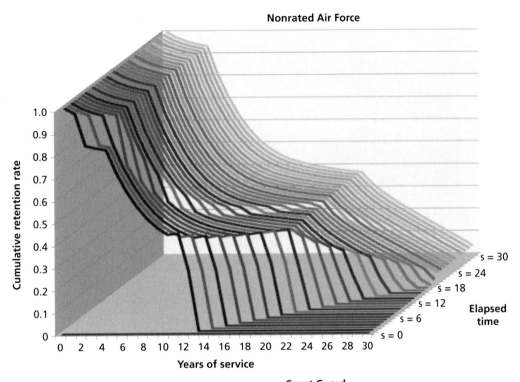

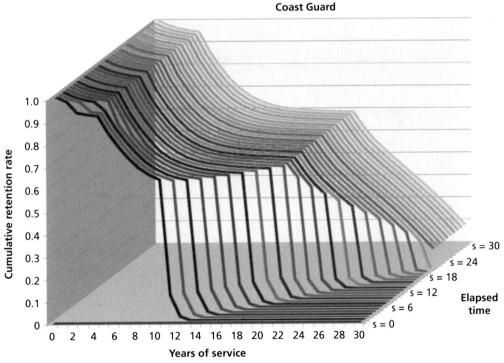

NOTE: s is the number of years since the policy change occurred.

Figure 5.6
Simulated Nonrated Air Force and Coast Guard Active Component Retention During the Transition for Officers Who Do Not Opt In, Optimized Continuation Pay

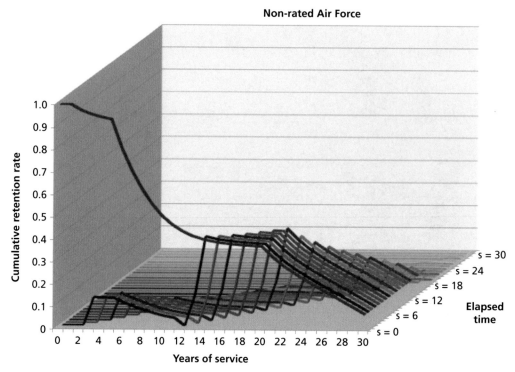

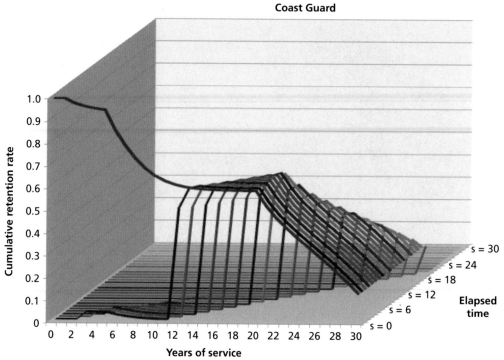

NOTE: s is the number of years since the policy change occurred.

**Figure 5.7
Simulated Retention During the Transition for Active Component Nonrated Air Force
Officers and for Coast Guard Officers, Minimum Continuation Pay**

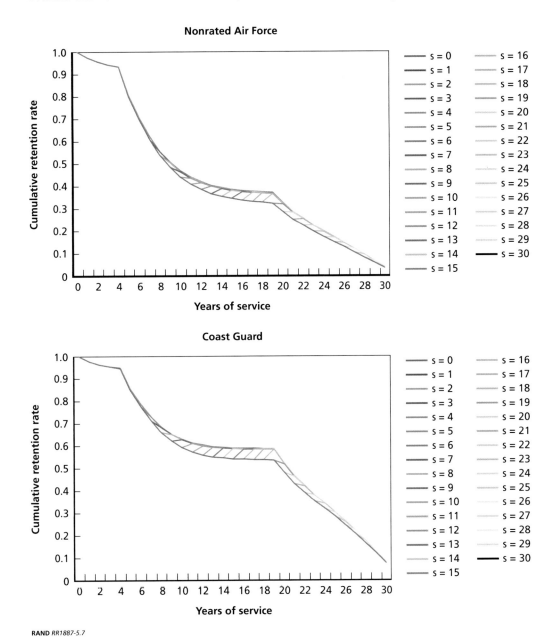

RAND *RR1887-5.7*

overall. In the second year, 2019, CP costs for Army officers reflect the higher opt-in rate of those who were in YOS 11 at the time of the opt-in decision in 2018 and who have 12 YOS in 2019, as well as Army retention between YOS 11 and YOS 12. As shown in Figure 5.3, the opt-in percentage is about 35 percent for Army officers at YOS 11. CP costs for Army officers jump up to $152 million in 2019. Enlisted CP costs for the Army also increase between 2018 and 2019 reflecting the increase in opt-in rates between YOS 12 and YOS 11, as well as Army retention between YOS 11 and YOS 12. Similar patterns are seen for the Navy and Marine Corps. The more gradual increase for the Air Force reflects gradual increase in opt-in rates between the YOS 12 cohort and the YOS 2 cohort for rated, as well as nonrated officers, shown in Figure 5.3.

Coast Guard CP costs increase from about $3 million to about $15.5 million between 2018 and 2019, reflecting the big increase in opt-in rates among officers between YOS 12 and YOS 11 shown in Figure 5.3. Enlisted CP costs grow more gradually, from about $2.5 million in 2018 to about $5 million in 2021 and eventually to $8.9 million by 2026. The more gradual growth is due to the gradual change in opt-in rates between the YOS 12 cohort and the YOS 2 cohort and the retention pattern between these YOS for Coast Guard enlisted personnel. As mentioned earlier, the gradual change in opt-in rates across YOS for the Coast Guard reflects the lower CP multiplier of enlisted personnel, the lower personal discount factor, and the missed TSP contributions in early career. Total CP costs for Coast Guard AC personnel increase sharply in the first year, from about $5.5 million to about $19 million, because of the sharp increase among officers, but then gradually increase to $24.7 million, reflecting the gradual growth in enlisted CP costs.

As mentioned earlier, we did not compute TSP costs and DB retirement costs. However, it is clear that, in the near term, costs will increase under the BRS for the Coast Guard. In addition to the CP costs shown in Figure 5.8, the Coast Guard will also be required to make TSP contributions on behalf of members, another source of cost. For the Air Force, Army, Marine Corps, and Navy, these increases in costs will be offset to some extent by any reductions in the accrual charge associated with the reduced DB benefit. However, Coast Guard retirement is not funded on an accrual basis but on a pay-as-you-go basis. As a result, Coast Guard DB retirement costs for AC personnel will not decrease until the retirement of service members under the BRS. Because only those with 12 or fewer YOS can elect the BRS, service members will not be reaching 20 YOS and eligibility to retire under the BRS and receive a DB annuity for at least eight years in the future. Thus, for at least eight years, and possibly longer, depending on opt-in behavior and TSP and CP costs, total costs will increase for the Coast Guard.

Figure 5.8
Blended Retirement System Active Component Continuation Pay Costs, by Year, Optimized Continuation Pay (in 2016 $M)

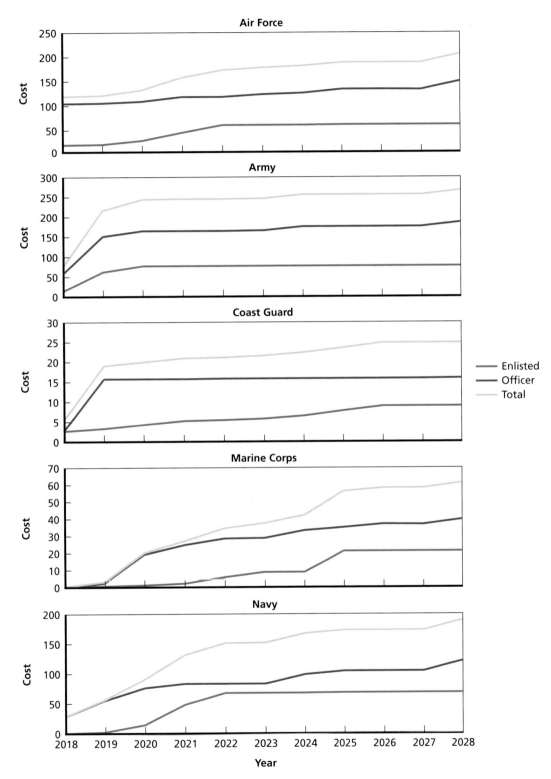

## CHAPTER SIX

Concluding Thoughts

Overview of Results

Our results show that the BRS can support a steady-state force and experience mix for each of the armed services that are close to that of their current forces, which were taken as the baseline. We assessed the retention effects of the BRS in terms of how well it could achieve the baseline. We find that the baseline force is achievable for enlisted personnel in each service when the CP multiplier is set at or near the floor for AC personnel of 2.5 and is achievable for officers when the CP multiplier is close to about one year's worth of basic pay—e.g., a multiplier of around 12 for each service. CP costs in the steady state increase AC costs for DoD by more than $700 million in 2016 dollars and for the Coast Guard by about $25 million. The BRS increases costs in the near term for the Coast Guard because retirement costs are funded on a pay-as-you-go basis and TSP contributions and CP add to costs. The DB plan for the Air Force, Army, Marine Corps, and Navy, however, is funded on an accrual basis. The accrual charge, which is determined by the DoD actuary, will be decreased in anticipation of lower future outlays for DBs, and the reduction in the accrual charge in the near term would help to offset the increased costs associated with CP.

Setting Continuation Pay Multipliers

The BRS reduces the DB retirement annuity by reducing the retirement benefit multiplier from 2.5 percent to 2.0 percent. The expected value of this reduction from the standpoint of service members is greater for officers in early and midcareers than for enlisted personnel for three reasons. First, officers have a higher likelihood than enlisted personnel of reaching 20 YOS. Second, they have higher basic pay than enlisted personnel at 20 YOS, so the reduced multiplier is applied to a larger number, implying a larger reduction in the DB annuity. Finally, we estimate a higher personal discount factor for officers than for enlisted personnel, so a given reduction in a future DB annuity is perceived as larger for officers. The addition of the TSP, which vests earlier than 20 YOS, partially offsets the reduced DB annuity, but the TSP benefit is not available until age 59.5. Consequently, the primary means available to the services for offsetting the reduced DB annuity and sustaining retention is CP.

A critical policy issue facing the services is the level at which to set CP multipliers. The BRS requires a minimum CP payment, but amounts above the minimum are left to the discretion of the services. We find that to sustain baseline retention, the officer CP multiplier should exceed the enlisted CP multiplier; for enlisted personnel, the optimized CP multiplier should be close to the minimum. The higher CP multiplier to sustain officer retention is due to the greater reduction for officers in the expected value of the DB annuity under the BRS.

While the analysis suggests different CP multipliers for officers and enlisted, the differences could raise concerns about equity between officers and enlisted personnel. After all, both officers and enlisted personnel have historically been under a common retirement system with no differentiation by rank. Insofar as CP is considered an element of the retirement system, CP could be considered an element that should not vary by rank. An alternative perspective is that officers and enlisted have different pay tables, and different CP multipliers would be consistent with this difference.

We used the DRM to explore the steady-state cost and retention effects of setting a common CP multiplier for officers and enlisted personnel. Thus, we considered a CP multiplier that is above 2.5 but below the roughly 12 months of basic pay for AC officers that we found would sustain officer retention. To illustrate the effects, we considered two cases: a CP multiplier of 5 and of 7, for both AC enlisted personnel and officers.[1] Figures 6.1–6.5 and 6.6–6.10 show the steady-state AC retention effects for enlisted personnel and officers for multipliers of 5 and 7, respectively, for each service. Table 6.1 extends Table 4.1; the final two columns show CP costs in the steady state for AC personnel when the CP multiplier is 5 and 7, respectively.[2]

As expected, setting the CP multiplier to a common level for AC personnel, above 2.5 but below the optimized CP multipliers for officers, results in enlisted retention above the baseline but officer retention below the baseline. These results assume the services do not take steps to curtail enlisted retention, such as tightening up-or-out rules.[3]

We find that CP costs are about the same or higher relative to the case of optimized CP multipliers. When the CP multiplier has a common level of 5, we estimate AC steady-state CP costs for the Coast Guard of $25.1 million as compared with $24.6 million for optimized multipliers. However, when the common CP multiplier is 7, we estimate CP costs of more than $35 million. For the DoD services, total CP costs for AC personnel in the Air Force, Army, Marine Corps, and Navy are $594.5 million when the CP multiplier is 5 and $851.2 million when it is 7. Yet these cost estimates for both DoD and for the Coast Guard understate the cost of the BRS because officer retention is not sustained. Additional special and incentive pays would be needed to sustain officer retention at baseline levels if the CP multiplier were set below a year's worth of basic pay. The table does not include those costs. The implication of the analysis is that setting a common CP multiplier for enlisted and officer personnel could address concerns about inequitable CP multipliers but would be inefficient relative to using optimized and different CP multipliers for officers and enlisted personnel.

---

[1]   For RC personnel, we assume the CP multiplier is 1 when it is 5 for AC personnel, and assume it is 2 when it is 7 for AC personnel.

[2]   For the Air Force, Figures 6.1 and 6.6 only show results for nonrated officers. The cost figures in Table 6.1 for the Air Force include CP costs for both rated and nonrated officers.

[3]   The services use up-or-out rules in managing their personnel force structure. Under an up-or-out rule, a service member must be promoted to a given grade within a certain number of YOS. Members not promoted may request a waiver, but it might not be granted.

Figure 6.1
Active Component Enlisted and Officer Personnel Under Legacy System Versus Blended Retirement System with Continuation Pay Multiplier = 5: Air Force

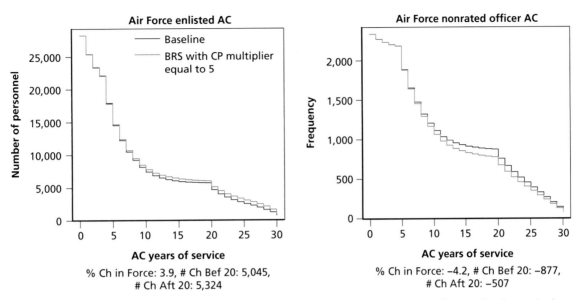

% Ch in Force: 3.9, # Ch Bef 20: 5,045,
# Ch Aft 20: 5,324

% Ch in Force: –4.2, # Ch Bef 20: –877,
# Ch Aft 20: –507

NOTE: "% Ch in Force" is the percentage change in the steady-state force size; "# Ch Bef 20" is the change in the steady-state number of personnel serving with fewer than 20 YOS; "# Ch Aft 20" is the change in the steady-state number of personnel serving with 20 or more YOS. These changes are relative to the 2009 force size, as shown in Table A.3.

RAND RR1887-6.1

Figure 6.2
Active Component Enlisted and Officer Personnel Under Legacy System Versus Blended Retirement System with Continuation Pay Multiplier = 5: Army

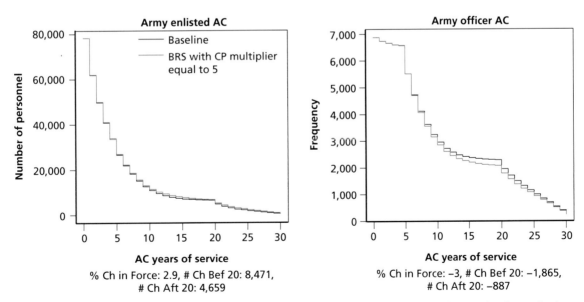

% Ch in Force: 2.9, # Ch Bef 20: 8,471,
# Ch Aft 20: 4,659

% Ch in Force: –3, # Ch Bef 20: –1,865,
# Ch Aft 20: –887

NOTE: "% Ch in Force" is the percentage change in the steady-state force size; "# Ch Bef 20" is the change in the steady-state number of personnel serving with fewer than 20 YOS; "# Ch Aft 20" is the change in the steady-state number of personnel serving with 20 or more YOS. These changes are relative to the 2009 force size, as shown in Table A.3.

RAND RR1887-6.2

Figure 6.3
Active Component Enlisted and Officer Personnel Under Legacy System Versus Blended Retirement System with Continuation Pay Multiplier = 5: Coast Guard

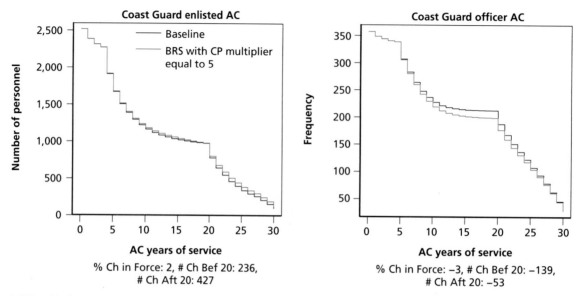

% Ch in Force: 2, # Ch Bef 20: 236,
# Ch Aft 20: 427

% Ch in Force: –3, # Ch Bef 20: –139,
# Ch Aft 20: –53

NOTE: "% Ch in Force" is the percentage change in the steady-state force size; "# Ch Bef 20" is the change in the steady-state number of personnel serving with fewer than 20 YOS; "# Ch Aft 20" is the change in the steady-state number of personnel serving with 20 or more YOS. These changes are relative to the 2009 force size, as shown in Table A.3.

RAND *RR1887-6.3*

Figure 6.4
Active Component Enlisted and Officer Personnel Under Legacy System Versus Blended Retirement System with Continuation Pay Multiplier = 5: Marine Corps

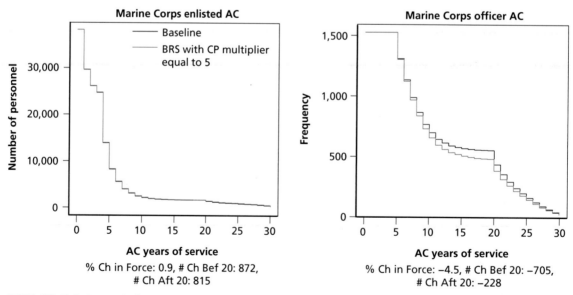

% Ch in Force: 0.9, # Ch Bef 20: 872,
# Ch Aft 20: 815

% Ch in Force: –4.5, # Ch Bef 20: –705,
# Ch Aft 20: –228

NOTE: "% Ch in Force" is the percentage change in the steady-state force size; "# Ch Bef 20" is the change in the steady-state number of personnel serving with fewer than 20 YOS; "# Ch Aft 20" is the change in the steady-state number of personnel serving with 20 or more YOS. These changes are relative to the 2009 force size, as shown in Table A.3.

RAND *RR1887-6.4*

Figure 6.5
Active Component Enlisted and Officer Personnel Under Legacy System Versus Blended Retirement System with Continuation Pay Multiplier = 5: Navy

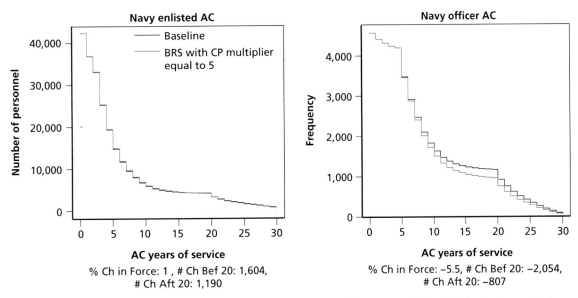

% Ch in Force: 1 , # Ch Bef 20: 1,604,
# Ch Aft 20: 1,190

% Ch in Force: −5.5, # Ch Bef 20: −2,054,
# Ch Aft 20: −807

NOTE: "% Ch in Force" is the percentage change in the steady-state force size; "# Ch Bef 20" is the change in the steady-state number of personnel serving with fewer than 20 YOS; "# Ch Aft 20" is the change in the steady-state number of personnel serving with 20 or more YOS. These changes are relative to the 2009 force size, as shown in Table A.3.

RAND RR1887-6.5

Figure 6.6
Enlisted and Officer Personnel Under Legacy System Versus Blended Retirement System with Continuation Pay Multiplier = 7: Air Force

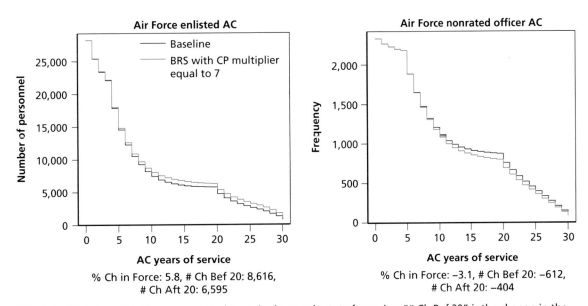

% Ch in Force: 5.8, # Ch Bef 20: 8,616,
# Ch Aft 20: 6,595

% Ch in Force: −3.1, # Ch Bef 20: −612,
# Ch Aft 20: −404

NOTE: "% Ch in Force" is the percentage change in the steady-state force size; "# Ch Bef 20" is the change in the steady-state number of personnel serving with fewer than 20 YOS; "# Ch Aft 20" is the change in the steady-state number of personnel serving with 20 or more YOS. These changes are relative to the 2009 force size, as shown in Table A.3.

RAND RR1887-6.6

Figure 6.7
Enlisted and Officer Personnel Under Legacy System Versus Blended Retirement System with Continuation Pay Multiplier = 7: Army

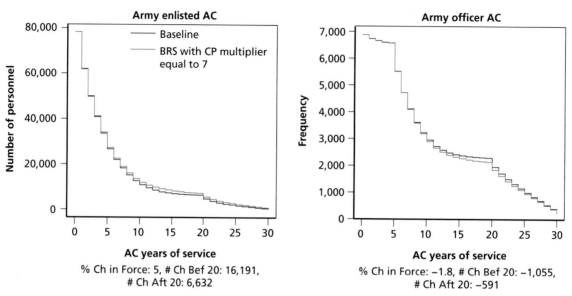

% Ch in Force: 5, # Ch Bef 20: 16,191,
# Ch Aft 20: 6,632

% Ch in Force: −1.8, # Ch Bef 20: −1,055,
# Ch Aft 20: −591

NOTE: "% Ch in Force" is the percentage change in the steady-state force size; "# Ch Bef 20" is the change in the steady-state number of personnel serving with fewer than 20 YOS; "# Ch Aft 20" is the change in the steady-state number of personnel serving with 20 or more YOS. These changes are relative to the 2009 force size, as shown in Table A.3.

RAND *RR1887-6.7*

Figure 6.8
Enlisted and Officer Personnel Under Legacy System Versus Blended Retirement System with Continuation Pay Multiplier = 7: Coast Guard

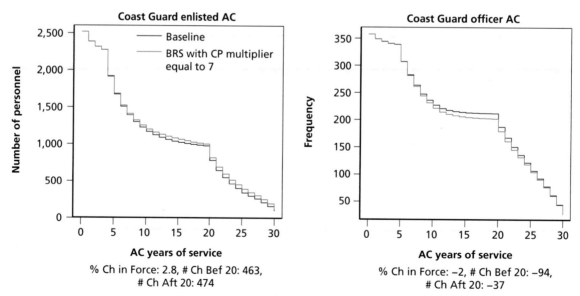

% Ch in Force: 2.8, # Ch Bef 20: 463,
# Ch Aft 20: 474

% Ch in Force: −2, # Ch Bef 20: −94,
# Ch Aft 20: −37

NOTE: "% Ch in Force" is the percentage change in the steady-state force size; "# Ch Bef 20" is the change in the steady-state number of personnel serving with fewer than 20 YOS; "# Ch Aft 20" is the change in the steady-state number of personnel serving with 20 or more YOS. These changes are relative to the 2009 force size, as shown in Table A.3.

RAND *RR1887-6.8*

Figure 6.9
Enlisted and Officer Personnel Under Legacy System Versus Blended Retirement System with Continuation Pay Multiplier = 7: Marine Corps

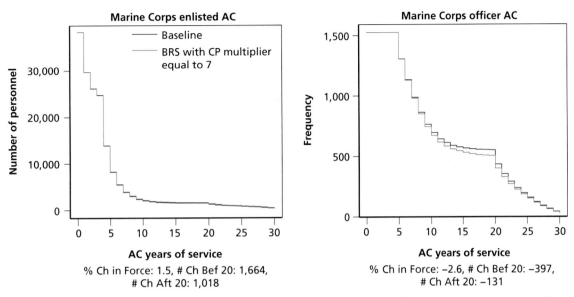

% Ch in Force: 1.5, # Ch Bef 20: 1,664,
# Ch Aft 20: 1,018

% Ch in Force: –2.6, # Ch Bef 20: –397,
# Ch Aft 20: –131

NOTE: "% Ch in Force" is the percentage change in the steady-state force size; "# Ch Bef 20" is the change in the steady-state number of personnel serving with fewer than 20 YOS; "# Ch Aft 20" is the change in the steady-state number of personnel serving with 20 or more YOS. These changes are relative to the 2009 force size, as shown in Table A.3.

RAND RR1887-6.9

Figure 6.10
Enlisted and Officer Personnel Under Legacy System Versus Blended Retirement System with Continuation Pay Multiplier = 7: Air Force

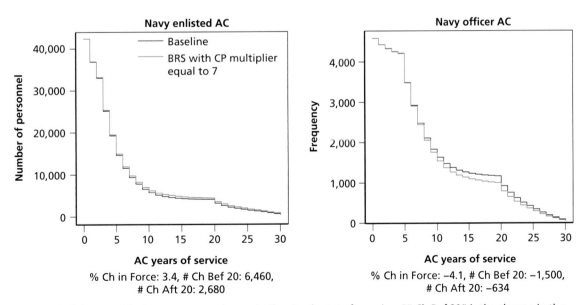

% Ch in Force: 3.4, # Ch Bef 20: 6,460,
# Ch Aft 20: 2,680

% Ch in Force: –4.1, # Ch Bef 20: –1,500,
# Ch Aft 20: –634

NOTE: "% Ch in Force" is the percentage change in the steady-state force size; "# Ch Bef 20" is the change in the steady-state number of personnel serving with fewer than 20 YOS; "# Ch Aft 20" is the change in the steady-state number of personnel serving with 20 or more YOS. These changes are relative to the 2009 force size, as shown in Table A.3.

RAND RR1887-6.10

Table 6.1
Steady-State Active Component Continuation Pay Costs Under Alternative Multipliers (in 2016 $M)

	Minimum CP Multiplier	Optimized CP Multiplier	CP Multiplier = 5	CP Multiplier = 7
Enlisted				
Air Force	56.5	56.5	113.3	161.5
Army	79.1	79.1	160.4	231.4
Coast Guard	8.8	8.8	18.4	25.9
Marine Corps	15.4	21.3	31.4	45.1
Navy	41.8	68.2	86.3	125.3
Officer				
Air Force	28.6	146.7	57.9	82.5
Army	41.2	186.0	82.5	116.4
Coast Guard	3.5	15.8	6.7	9.5
Marine Corps	9.6	39.5	19.2	27.3
Navy	21.6	120.8	43.5	61.7
Total cost				
Air Force	85.0	203.1	171.20	244.00
Army	120.3	265.1	242.90	347.80
Coast Guard	12.3	24.6	25.10	35.40
Marine Corps	25.0	60.9	50.60	72.40
Navy	63.4	189.0	129.80	187.00

Continuation Pay Flexibility

Some services might prefer to have CP multipliers vary by occupational area or skill area, and the services might want to vary when CP is paid across skill areas (e.g., YOS eight or ten versus 12). Currently we are not able to model variability across skill areas. CP policy allows for such variability and the services might want to pursue this if requirements permanently change for some skills but not others and the change is not easily handled with traditional special and incentive pay. Additional flexibility in how CP is paid could result in some increase in efficiency; however, we are not able to quantify how large this potential increase might be with our current simulation capability.

Final Thoughts

To assess the effects of the BRS on retention and cost for the Air Force, Army, Marine Corps, and Navy, we used estimates from our active and reserve DRM for enlisted personnel and officers and developed computer code to simulate the BRS effects developed in past research.

While past RAND reports used the DRM to provide analysis of proposals leading up to the BRS, DoD lacked information on the retention and cost effects of the BRS specifically. In the case of the Coast Guard, we developed a new longitudinal database of retention over enlisted and officer careers for this project and estimated models, similar to models we developed for the Air Force, Army, Marine Corps, and Navy.

The modeling capability for all five armed services could be used to assess additional changes to the BRS under consideration, such as allowing CP to be paid at other YOS, e.g. YOS eight to 12, as well as the retention effects of the BRS for members who choose the lump-sum option, given an assumed discount rate for computing the lump sum. Beyond changes to the retirement system, the DRM for the Air Force, Army, Marine Corps, and Navy has been used to consider the effects of other compensation policies on retention, such as alternative changes to the basic pay table and the introduction of separation incentive pay to facilitate a drawdown (Asch, Hosek, Kavanagh, and Mattock [2016]; Mattock, Hosek, and Asch [2016]). Like the DRM capability for the other services, the new DRM capability for the Coast Guard could have future applicability to a range of compensation policies.

The Dynamic Retention Model

The DRM is an econometric model of officer and enlisted behavior. It models service members as being rational and forward-looking, taking into account both their own preference for military service and uncertainty about future events that may cause them to value military service more or less, relative to civilian life. At each decision point in an active-duty career, the individual compares the value of leaving the military with the value of staying, taking into account that the decision to stay can be revisited at a later time.

This appendix describes the model in general and discusses the specific application to the Coast Guard, including data and model estimates. As mentioned in the main text, we relied on DRM estimates for the Air Force, Army, Marine Corps, and Navy produced for the 11th QRMC and documented elsewhere, but we estimated new DRMs for the Coast Guard. Because enlisted Coast Guard retention has varied substantially over different entry cohorts from 1990 to 2015, we extended the DRM for Coast Guard enlisted personnel to capture this feature. Retention for Coast Guard officers shows less variability, and we were able to pool data across entry cohorts, as we have done in the past for the other services. We further discuss this issue in this appendix and how we extended the DRM to account for it.

The Behavioral Model Underlying the Dynamic Retention Model

The behavioral model underlying the DRM is conceptually simple. During each period of active service, the individual compares the value of staying in the AC with leaving and bases his or her decision on which alternative has the maximum value. Every year after leaving active service, a member can compare the value of participating in the RC with the value of leading a purely civilian life and choose the alternative that yields the maximum value for that year.

Although this model is relatively simple, the implications can be intricate, because an individual can choose to revisit the decision to stay in the AC or participate in RC service at a later date, and that decision will depend on his or her unique circumstances at a given point in time. Those circumstances include relative preference for AC or RC service to a purely civilian life and random events that may affect relative preferences over AC, civilian, and RC alternatives.

In the model, the value of staying depends upon the individual's preference for active military service (or "taste" for active service, which is assumed to be constant over time), the compensation received for active service, the expected maximum of the value of staying and leaving in the next period, and a period- and individual-specific environmental disturbance term (or "shock") that can either positively or negatively affect the value placed on active service in that period. For example, an unusually good assignment would increase one's relative preference for active service, while having an ailing family member who requires assistance with home care may decrease the value placed on active service. The value of staying also includes the value of the option to leave at a later date, that is, the individual knows that he or she can revisit the decision to stay the next time it is possible to make a retention decision.

We make the simplifying assumption that individuals do not reenter once they have left active service. While there are instances in which people do reenter the AC, the vast majority of those who leave do not reenter. This assumption substantially reduces the number of possible career paths that need to be evaluated and makes the model more tractable.

An individual who leaves the AC can choose to either be a civilian or combine civilian life with RC service. A person can join the RC immediately after leaving active service or can choose to join at a later date. Once a person enters the RC, he or she is free to choose to stay or to leave with the option of reentering at a later date, service regulations permitting.

At the beginning of each year, RC members compare the value of the civilian alternative— that is, leading a purely civilian life for that year—with the value of the RC alternative—that is, a first or additional year of RC service—and choose the alternative that yields the maximum value.

The value of the civilian alternative includes the civilian wage, the AC or RC military retirement benefit that the individual is entitled to receive (if any), an individual- and period-specific shock term that can either positively or negatively affect preference for the civilian alternative, and the future option to enter (or reenter) the RC, service regulations permitting.

The value of RC service includes the civilian wage, the RC compensation to which the individual is entitled, given his or her cumulative AC and RC service, an individual- and period-specific shock term that can either positively or negatively affect the preference for the RC alternative, and the future option to either continue in the RC or return to a purely civilian life.

Technical Details

In each period, the active service member compares the value of staying in the AC with the value of leaving and joining the RC or entering civilian life. We use a nested logit approach to capture this decision, where the active service member is modeled as comparing active service with a civilian/RC nest.

Active service has the value

$$V_a + \epsilon_a ,$$

where V_a is the nonstochastic portion of the value of the active alternative, and ϵ_a is the environmental disturbance (shock) term specific to the active alternative, assumed to be extreme-value distributed.

The civilian/RC nest has the value

$$\max \left[V_r + \omega_r, V_c + \omega_c \right] + \upsilon_{rc} ,$$

where V_r is the nonstochastic portion of the value of the RC alternative; V_c is the nonstochastic portion of the value of the civilian alternative; ω_r and ω_c are the shock terms specific to the RC and civilian alternatives, respectively; and v_{rc} is the civilian/RC nest-specific shock.

The mathematical symbols for nonstochastic values and shock terms are summarized in Table A.1.

The value of staying in the AC is the sum of the individual's taste for active service, γ_a; active military compensation, ω_a; and the discounted value of the expected value of the maximum of the AC, civilian, and RC alternatives in the following period. Note that to calculate wages, eligibility for retirement benefits, and so on, we need to keep track of time spent in the AC, time in the RC, and time overall. Thus, we have three time indexes that are each incremented appropriately to reflect the result of the choice in the current period. For example, if an individual serves a year in the AC, then both the time in the AC and the total time will be incremented by one:

$$
\begin{aligned}
&V_a(t_{active}, t_{reserve}, t_{total}) \\
&= \gamma_a + W_a(t_{active}) \\
&+ \beta E[\max[V_a(t_{active}+1, t_{reserve}, t_{total}+1) \\
&+ \in_a, \max[V_r(t_{active}+1, t_{reserve}, t_{total}+1) + \omega_r, \\
&V_c(t_{active}+1, t_{reserve}, t_{total}+1) + \omega_c]]]
\end{aligned}
$$

The value of the RC alternative is the sum of the individual's taste for RC service, γ_r; RC military compensation, W_r; civilian compensation, W_c; and the discounted value of the expected value of the maximum of the civilian and RC alternatives in the following period:

Table A.1
Mathematical Symbols for Nonstochastic Values and Shock Terms

Symbol	Interpretation
V_a	Nonstochastic value of the AC alternative
V_r	Nonstochastic value of the RC alternative
V_c	Nonstochastic value of the civilian alternative
ϵ_a	Active alternative specific shock term, $\in_a \sim EV[0, \sqrt{\lambda^2 + \tau^2}]$
ω_r	RC alternative specific shock term, $\omega_r \sim EV[0, \lambda]$
ω_c	Civilian alternative specific shock term, $\omega_c \sim EV[0, \lambda]$
v_{rc}	Civilian/RC nest-specific shock term, $v_{rc} \sim EV[0, \tau]$

$$V_r(t_{active}, t_{reserve}, t_{total})$$
$$= \gamma_r + W_c(t_{total}) + W_r(t_{active}, t_{reserve})$$
$$+ \beta E[\max[V_r(t_{active}, t_{reserve} + 1, t_{total} + 1) + \omega_r,$$
$$V_c(t_{active}, t_{reserve} + 1, t_{total} + 1) + \omega_c]]$$

Finally, the value of the civilian alternative is the sum of civilian compensation, W_c; any active or RC service retirement benefit that the individual is eligible for, R; and the discounted value of the expected value of the maximum of the civilian and RC alternatives in the following period

$$V_c(t_{active}, t_{reserve}, t_{total})$$
$$= W_c(t_{total}) + R(t_{active}, t_{reserve}, t_{total})$$
$$+ \beta E[\max[V_r(t_{active}, t_{reserve}, t_{total} + 1) + \omega_r, V_c(t_{active}, t_{reserve}, t_{total} + 1)$$
$$+ \omega_c]]$$

The mathematical symbols for taste and compensation are summarized in Table A.2.

We also assume that individuals' tastes for AC and RC service are bivariate normally distributed. Given these distributional assumptions, we can derive choice probabilities for each alternative and write an appropriate likelihood equation to estimate the parameters of the model (the parameters of the probability distribution for the shock terms; the population distribution of taste for AC and RC service, consisting of the means and standard deviations of AC and RC taste, as well as the correlation between taste for AC service and RC service; the discount factor; and switching costs associated with leaving AC service for civilian life or entering the RC from civilian life). These derivations are documented in Asch et al. (2008). The model is estimated by maximum likelihood. Optimization is done using the Broyden-Fletcher-Goldfarb-Shanno (BFGS) algorithm, a standard hill-climbing method. Standard errors of the

Table A.2
Mathematical Symbols for Taste and Compensation

Symbol	Interpretation
γ_a	Taste for active service relative to civilian alternative, $\{\gamma_a, \gamma_r\} \sim N[M, \Sigma]$
γ_r	Taste for RC service relative to civilian alternative, $\{\gamma_a, \gamma_r\} \sim N[M, \Sigma]$
W_a	AC compensation (regular military compensation [RMC])
W_c	Civilian compensation
W_r	RC compensation
β	Discount factor
R	Military retirement benefit

estimates were computed using numerical differentiation of the likelihood function and taking the square root of the absolute value of the diagonal of the inverse of the Hessian matrix.

Model Extension for Coast Guard Enlisted Personnel

Coast Guard enlisted retention has varied substantially over time, and this poses a unique challenge requiring us to extend the DRM for enlisted personnel to allow for the variation. Fewer than half of the enlisted members in entry cohorts from 1990 to 1998 served five or more years. In contrast, 50 percent to 65 percent of the members in the entry cohorts from 1999 to 2003 served for five or more years. The percentage of enlisted members serving eight or more years doubled from around 20 percent in early 1990s cohorts to more than 40 percent in cohorts entering after 2000. In addition, Coast Guard enlisted end strength (end-of-year inventory) has also varied over time, from just under 24,000 in the year 2000 to more than 31,000 in 2012. Thus, both the shape and the size of the Coast Guard enlisted force have changed over the past two decades. As a result the new force is larger and more experienced.

This means that the enlisted retention curve has shifted up over time, with earlier cohorts showing greater attrition in the early YOS and more recent cohorts showing less attrition. This can be seen in Figure A.1, which shows the cumulative retention rate through 2014 for each entering cohort from 1990 to 2003. In contrast, Coast Guard officer retention curves have been more consistent over time. Because of this relative consistency over time for officers, we can pool data across entry cohorts and estimate the officer DRM with the pooled data. The results shown below for officers are based on pooled data.

However, pooling is not feasible for Coast Guard enlisted personnel because of the variance over time in the retention profiles. This variance can pose a problem in estimating a model of enlisted Coast Guard. We would like to take advantage, to the extent that is possible, of all the historical data available when estimating a retention model because earlier cohorts give us insight into how people make decisions throughout their careers, whereas more recent cohorts can give us insight into only the early part of a career. Mid- through late-career data are particularly valuable, as these tend to be the years when the possibility of collecting the DB retirement annuity at 20 YOS weighs most heavily on people's minds. While the retention experience of more-recent cohorts may be closer to what we expect to see in the near future for the Coast Guard, recent cohort data alone do not cover enough YOS to allow us to confidently estimate the retention model parameters.

Simply pooling the data across cohorts, as we did for officers, is not a satisfactory approach, as we demonstrate in Figure A.2. The figure shows the retention profile when we pool the 1994 to 2007 data. The resulting retention profile falls short of recent Coast Guard year-to-year retention rates.

The approach we took to address this variability problem was to allow value of staying, V_a, to vary over time as a function of enlisted end strength, $S(year)$, where *year* refers to calendar year. The underlying idea is that the attractiveness of staying in the Coast Guard will vary positively with the end strength, as this will mean that there are more opportunities in higher years of service for enlisted personnel. End strength enters into the value of staying as an additive term multiplied by the coefficient δ. Given an individual's entry cohort and YOS, we can calculate the calendar year and get the end for that year, as $S(year) = S(cohort + t_{active} - 1)$. The value of staying for enlisted members then becomes:

Figure A.1
Enlisted Cumulative Retention Rate, by Year Cohort Entered Active Service

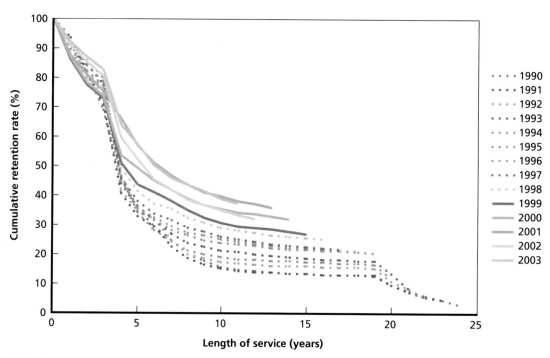

NOTE: Cohorts with fewer than half of the members serving five or more years are shown using dashed lines, while later cohorts with more than half of the members serving five or more years are shown using solid lines.

RAND RR1887-A.1

Figure A.2
Pooled Longitudinal Data Do Not Resemble Recent Coast Guard Retention Experience

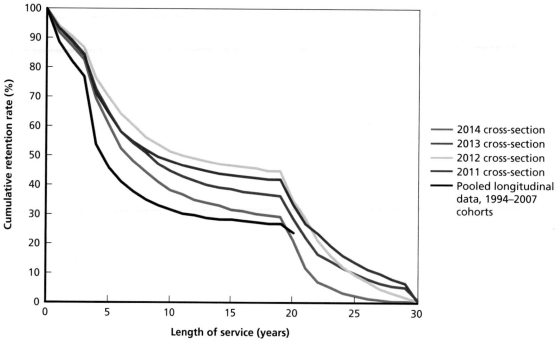

RAND RR1887-A.2

$$V_a(t_{active}, t_{reserve}, t_{total})$$
$$= \gamma_a + \delta S(cohort + t_{active} - 1) + W_a(t_{active})$$
$$+ \beta E[\max[V_a(t_{active} + 1, t_{reserve}, t_{total} + 1)$$
$$+ \in_a, \max[V_a(t_{active} + 1, t_{reserve}, t_{total} + 1)$$
$$+ \omega_r, V_c(t_{active} + 1, t_{reserve}, t_{total} + 1) + \omega_c]]]$$

This approach works well, as can be seen in the fit charts shown later in this appendix, in Figure A.3 for enlisted and Figure A.5 for officers. Furthermore, the estimated coefficient on end strength is positive and significant, as we would expect.

Data

For our models for the Air Force, Army, Marine Corps, and Navy, our main data file is the Work Experience File (WEX). The WEX contains person-specific longitudinal records of active and reserve service.[1] The Defense Manpower Data Center (DMDC) creates WEX data from the active-duty master file and the RC common personnel data system file. DMDC uses these files to build a snapshot of all personnel for each reporting period that includes demographic and work experience information. To maintain the file, DMDC compares data for the current and previous periods and creates three types of records—a gain record, a loss record, and a change record. A gain record is created when a service member's Social Security number is not in the previous period but is in the current one. A loss record is created when a Social Security number appears in the previous period but not in the current one. When a loss occurs, all related work experience records are moved to a loss file; these are retrieved when an individual reenters service; i.e., when a gain occurs. A change record is created when there is a change in any of seven variables: service/component, pay grade, reserve category code, primary service occupation code, secondary service occupation code, duty service occupation code, and unit identification code. The WEX record also includes a member's age and gender.

The WEX data created by DMDC begin with service members in the AC or RC on or after September 30, 1990. Our analysis file includes AC non-prior-service entrants in 1990 and 1991 followed through 2010, providing 21 years of data on 1990 entrants and 20 years on 1991 entrants. For each AC, we drew samples of 25,000 individuals who entered the component in FY 1990–1991, constructed each service member's history of AC and RC participation, and used these records in estimating the model. We used WEX variables to identify an individual's component and branch of service (e.g., AC Army, RC Army Reserve) by year from the date of entry onward. We used pay entry base date and component/branch in counting years of AC service and years of RC participation following AC service.

We constructed samples for each service and for enlisted personnel and officers. In constructing the officer samples, we excluded medical personnel and members of the legal and chaplain corps because their career patterns differ markedly from those of the rest of the officer corps. Analysis of retention for these personnel needs to be conducted separately. For a similar reason, for Air Force officers, we excluded rated pilots.

---

[1] WEX is used primarily for production of Verification of Military Training and Experience DD Form 2586 documents.

The WEX does not include Coast Guard personnel, so we built WEX-like longitudinal files from individual snapshots of all Coast Guard personnel present in the AC and RC in each quarter from 1990 to 2015. The data we used are quarterly master active-duty and master reserve-duty files provided by the DMDC.

As with the WEX, the longitudinal file for the Coast Guard begins with service members who begin the AC with no prior AC or RC service, and we track individuals over the course of the AC enlisted career or AC officer career, and, if they separate, over their career in the Selected Reserves. We excluded individuals who began as an enlisted member and become either a warrant officer or commissioned officer during their careers, because modeling the decision to become a warrant or commissioned officer was beyond the scope of our study. We used variables to identify an individual's component (e.g., AC or RC) and enlisted or officer status by year from the date of entry onward. We used pay entry base date and component/ branch in counting years of AC service and years of RC participation following AC service.

The other key source of data for all five services was information on military and civilian pay. AC pay, RC pay, and civilian pay are averages based on the individual's years of AC, RC, and total experience, respectively. AC and RC pay are also related to military retirement benefits. Annual military pay for AC members is represented by RMC for FY 2009, equal to the sum of basic pay, basic allowance for subsistence (BAS), basic allowance for housing (BAH), and the federal tax saved because the allowances are not taxed. We computed RMC by year of service for enlisted and officers using the RMC tables provided by OSD and weighting them with the 2009 grade-by-YOS inventory of enlisted and officer personnel in each service.

RC members are paid differently from AC members, although the same pay tables are used. Reservists who are drilling but not on active duty receive subsistence allowance for their two drilling days per month and do not receive a housing allowance. Reservists on active-duty training receive rations and housing in kind during the two weeks of training and receive either a partial housing allowance or a rate applied for married members, unless they are housed in contract housing off-base.

RC pay is based on years of AC service and years of RC participation. We average it over pay grade and dependents status using RC strength information from the 2007 report *Official Guard and Reserve Manpower Strengths and Statistics* (Office of the Assistant Secretary of Defense, Reserve Affairs [2007]). Reserve pay in a year is calculated as the sum of drill pay for four drills per month, 12 times a year, plus pay for 14 days of active-duty training, typically done in the summer. Drill pay is 1/30 of monthly basic pay for each drill period, or 4/30 per weekend. During each day of active-duty training, the reservist receives basic pay plus BAS. Single members receive BAH for a service member without dependents, while married members receive BAH for a service member with dependents. In our calculation, RC members receive BAH reserve component/transit (RC/T), a housing allowance for certain circumstances, including being on active duty fewer than 30 days. Given years of service and grade, we compute a reservist's annual pay as

12 × weekend drill pay + 14 × [(BAS + daily basic pay) + %married × BAH RC/T for those with dependents + %single × %on base × BAH RC/T for those without dependents) + tax advantage]

To incorporate the tax advantage, we use the same adjustment as for AC annual pay, 6 percent. Some reservists receive special and incentive pay such as bonuses, but these are not included explicitly. Instead, their role is generally picked up in the reserve taste term. Also, the model does not address the activation and deployment of reservists.

The reserve retirement benefit formula and the active-duty high-three retirement formula are programmed into our model. Regarding the computation of reserve retirement benefits, we assume an RC participant accumulates 75 points per year. Unlike AC retirement benefits, which start as soon as the AC member retires from service, RC retirement benefits begin at age 60.[2] The formula for RC retirement benefits under the legacy system is the same as that for AC retirement benefits, with the proviso that RC retirement points are converted into years of service (for the purpose of retirement) by dividing total points by 360. A year of AC service counts as a full year. Reservists who qualify for reserve retirement benefits can transfer to the "retired reserve," which means that their high-three pay is based on the basic pay table in place on their 60th birthday, and their basic pay is based on their pay grade and years in grade, in which the latter include years in the retired reserve.[3]

Civilian Earnings

For enlisted personnel, civilian earnings are the 2007 median wage by experience for full-time male workers with an associate's degree. For officers, we use the 2007 80th percentile of earnings for full-time male workers with a master's degree in management occupations. The data are from the U.S. Census Bureau (DeNavas-Walt, Proctor, and Smith [2008]). Civilian work experience is defined as the sum of active years, reserve years, and civilian years since age 20; however, in this report, it does not vary by other such factors as years since leaving active duty.

Approach to Simulation

To simulate retention behavior, we first create a synthetic population of 10,000 individuals entering active duty by randomly drawing tastes from the estimated AC and RC taste distribution. This synthetic population is large enough for the simulations to produce AC retention and RC participation careers that, when aggregated, provide AC cumulative retention and RC participation curves representative of the policy being simulated. Each pair of AC and RC taste draws represents an individual entering active duty. We also draw shocks for each year for each synthetic individual from the shock distributions. We assume that the synthetic individuals follow the logic of the model, and we specify the compensation policy for the simulation. We simulate behavior under the legacy retirement system, the baseline, and then simulate it under the BRS. The simulations produce a 30-year record of AC retention and RC participation for each member of the synthetic population under each retirement policy.

We use the data sets of simulated behavior to tabulate AC and RC retention and, along with information on compensation, to compute policy cost. In the case of the BRS, the policy

---

2   If the RC member has been deployed in the period beginning on January 28, 2008, retirement age is decreased by three months for every 90 consecutive days of deployment. This change is not included in our model because the model does not include deployment.

3   In addition, military retirees (including reserve retirees receiving retired pay) are eligible to receive health care through TRICARE for the remainder of their lives, as can their spouses, and coverage continues for the spouse if the retiree dies and she or he does not remarry. "Gray area" retirees (i.e., members of the retired reserve who are not drawing retired pay) may purchase TRICARE coverage under the TRICARE Retired Reserve program until they become eligible for TRICARE. However, we did not model the health benefit.

cost we compute is CP cost. The simulation outputs include graphs of AC retention by year of service and RC participation by year of active-plus-reserve service, as well as tabulations for AC and RC of force size and CP costs.

Under the assumption of a steady state, the AC force size of the simulated population is the count of individuals present in each year up to year 30. This count is scaled up to the 2009 AC force size for each service shown in Table A.3.

RC force size is based on the count of simulated individuals participating in the RC at each year of service, given scaling up the AC force to its force size. As mentioned, RC YOS are based on the number of active years plus reserve years.[4]

Simulation of Continuation Pay Cost

CP costs for the AC equal CP costs at YOS 12 multiplied by the number of AC personnel at YOS 12. CP costs are scaled up to the 2009 force size and expressed in 2016 dollars.

Table A.3
2009 Active Component Force Size, for Scaling Simulations

Service	Force Size
Air Force	
Enlisted	263,351
Officer	65,496
Army	
Enlisted	458,220
Officer	90,795
Coast Guard	
Enlisted	33,172
Officer	6,421
Marine Corps	
Enlisted	182,366
Officer	20,709
Navy	
Enlisted	272,208
Officer	52,031

---

[4] As an example of this count, consider someone who over the course of 40 years (ages 20 to 60) had five years of AC and five years of RC service. This individual would be present in the RC at six YOS (5 + 1), 7 (5 + 2), 8, 9, and 10 (participation in the RC could have occurred in nonconsecutive calendar years). In each of these years, the individual would be counted in the steady-state RC force. Because everyone begins in the AC, the smallest RC YOS entry is 2 (1 + 1).

Estimation and Coefficient Estimates for Coast Guard Enlisted and Officer Personnel

The parameters we estimate fall into several sets. The first is the set of parameters related to the probability distributions of the shock terms, τ and λ. The second is the set related to the population distribution of AC and RC taste, which is assumed to be bivariate normal. The mean of AC taste is μ_1 and the standard deviation is σ_1; the mean of RC taste is μ_2 and the standard deviation is σ_2; and the correlation between active and reserve taste is ρ.

The third is the set of switching costs. The enlisted model has three switching costs: Switch1, the cost associated with leaving AC service before the end of the initial term of service; Switch2, the cost associated with entering the RC from civilian life; and Switch3, the cost of entering the RC from the AC after completing the initial term of service.

The officer model has two switching costs, Switch1, the cost associated with leaving AC service before the end of the initial term of service, and Switch2, the cost associated with entering the RC from civilian life. Exploratory analysis showed that Switch3 was not needed in the officer model.

The switching costs enter the appropriate value functions as an additive term to the instantaneous utility received in that period. For example, Switch1, the cost associated with leaving AC service before the end of the initial term of service, enters V_r, the value of RC service, as an additive term to the instantaneous utility received in the period the member enters the RC if the member was in the AC the prior period and if the member left AC service *before* completing his or her initial term of service. Switch1 also enters V_c, the value of a purely civilian life, in the same way.

Switch2 and Switch3 both enter only into the value of RC service. Switch2 enters V_r as an additive term to the instantaneous utility received in the period the member enters the RC if the member was in the AC the prior period and if the member left AC service *after* completing his or her initial term of service. Similarly, Switch3 enters V_r in the period the member enters (or reenters) the RC if the member was a civilian in the prior period.

The fourth set applies only to the enlisted model, and it is the coefficient δ on end strength for that year measured in thousands, as discussed earlier. Finally, the discount factor β is not estimated but is set at a value consistent with the values estimated for enlisted personnel and officers in earlier work. These values are 0.88 for enlisted and 0.94 for officers.

All coefficients are estimated in logarithms (multiplying by negative one as appropriate), with the sole exception of ρ. The correlation coefficient is measured as the inverse hyperbolic tangent, as the hyperbolic tangent function is a convenient way of mapping the real line to a [–1,1] interval. The back-transformed coefficients are reported in the "actual" column of the tables; coefficients corresponding to monetary values are measured in thousands of dollars.

We use maximum likelihood to estimate the coefficients for the DRM, and as a result we can test the coefficients for statistical significance. The development of the likelihood function and the parameter estimation procedure are documented in detail in Mattock, Hosek, and Asch (2012). In a nutshell, the likelihood function describes the probability of a given career, consisting of a sequence of active, civilian, and reserve states. To compute these probabilities we must compute the probability of choosing each alternative in each time period. The dynamic program can be solved for given values for active and reserve taste, and, given our assumption of an extreme-value distribution for the shock terms, the solution has a closed-form for the probability of choosing each of the two or three alternatives available at any given time. As mentioned, we use these to construct a career probability for each individual. The expression for the career probability implicitly depends on the parameters to be estimated—e.g., mean active taste, mean reserve taste, discount rate, and so forth. Because tastes are not known at the individual level, we numerically integrate out heterogeneity in taste.[5] Numerical optimization is done using the BFGS algorithm, a standard hill-climbing method.

Standard errors are computed using numerical differentiation of the likelihood function at the parameter estimates to produce the matrix of second derivatives, the Hessian matrix. The standard errors are computed using the square root of the absolute values of the diagonal of the inverse of the Hessian.

Coast Guard Parameter Estimates

We report the parameter estimates and standard errors by service for enlisted and officers in Tables A.4 and A.5, respectively. The final column in the tables shows the transformed coefficient estimates. As mentioned, to make the numerical optimization easier, we did not estimate most of the parameters directly, but instead estimated the logarithm of the absolute value of each parameter (except for the taste correlation where we estimated the inverse hyperbolic tangent of the parameter). To recover the parameter estimates, we transformed the estimates back to their unlogged values.

All the coefficients in the Coast Guard enlisted model are statistically significant and show the expected signs. The within-nest shock term for RC and civilian alternatives, λ, at

Table A.4
Coast Guard Enlisted Dynamic Retention Model Estimates, Cohorts 1994–2003, Beta = 0.88

Parameter	Coefficient Estimate	Standard Error	Z-Statistic	Transformed Value
Tau (τ)	3.6873	0.1545	23.8653	39.9353
Lambda (λ)	3.2546	0.2146	15.1673	25.9098
Mu1 (μ_1)	2.7938	0.0524	53.3617	−16.3436
Mu2 (μ_2)	4.5957	0.2673	17.1962	−99.0601
SD11 (σ_1)	3.0588	0.0845	36.1886	21.3026
SD22 (σ_2)	4.1726	0.2736	15.2495	64.8817
atanhRho (ρ)	0.5581	0.0098	56.9959	0.5066
Switch1	4.7881	0.0729	65.6633	−120.0717
Switch2	4.8436	0.2115	22.8960	−126.9221
Switch3	1.5753	0.3856	4.0853	−4.8322
Delta (δ)	1.1994	0.0764	15.6964	3.3182

NOTE: Transformed parameters are denominated in thousands of dollars, with the exception of the taste correlation.

---

[5] For trial values of the taste distribution parameters, possible tastes for the individual are drawn from the distribution. For each taste draw, the career probability expression is evaluated and an average of those probabilities is taken, where the weight on a probability depends on the probability of the drawn taste.

Table A.5
Coast Guard Officer Dynamic Retention Model Estimates, Cohorts 1990–2007, Beta = 0.94

Parameter	Coefficient Estimate	Standard Error	Z-Statistic	Transformed Value
Tau (τ)	4.6036	0.1015	45.3534	99.8420
Lambda (λ)	2.9966	0.2059	14.5567	20.0170
Mu1 (μ_1)	2.1393	0.2158	9.9142	–8.4931
Mu2 (μ_2)	4.6621	0.2120	21.9876	–105.8552
SD11 (σ_1)	2.9459	0.2785	10.5781	19.0270
SD22 (σ_2)	3.8988	0.2524	15.4466	49.3425
atanhRho (ρ)	–0.6489	0.0460	–14.0944	–0.5709
Switch1	5.6590	0.1134	49.9159	–286.8545
Switch2	4.5106	0.2113	21.3507	–90.9776

NOTE: Transformed parameters are denominated in thousands of dollars, with the exception of the taste correlation.

$25,909 is smaller than the between-nest shock term, τ, at $39,935, reflecting the difference between active duty and reserve or civilian life. The mean tastes for AC service and RC service are both negative and significant, at –$16,343 and –$99,060, respectively; the large negative coefficient for RC service may be due to the very small size of the Coast Guard Reserve. The standard deviations are both positive and significant, at $21,302 and $64,881 for AC and RC taste, respectively; tastes for the AC and the RC are positively correlated at 0.50. The switching cost associated with leaving the AC before the end of the initial term is substantial, at –$120,071. The cost associated with joining the RC after spending some time as a civilian is also substantial, at –$126,922, whereas the cost associated with going directly from the AC to RC service is significantly less, at –$4,832. Finally, the coefficient on enlisted end strength is $3,318, meaning that an increase of 1,000 enlisted members is equivalent to an increase in the nonpecuniary value of staying by $3,318.

In the officer model, the within-nest shock term for RC and civilian alternatives, λ, at $20,017 is smaller than the between-nest shock term, τ, at $99,842. The mean tastes for AC service and RC service are both negative and significant at –$8,493 and –$105,855, respectively; as we noted when discussing the enlisted coefficients, the large negative coefficient for RC service may be due to the very small size of the Coast Guard Reserve. The standard deviations are both positive and significant, at $19,027 and $49,342 for AC and RC taste, respectively, showing a pattern similar to that seen in the enlisted population. The tastes for the AC and the RC are negatively correlated at –0.57; this is atypical for service-level DRMs of officer behavior. However, a negative correlation is something we have seen in DRMs of military mental health care providers. The switching cost associated with leaving the AC before the end of the initial term is even more substantial than what we see for Coast Guard enlisted, at –$286,854. The cost associated with joining the RC after spending time as a civilian is –$90,977; most RC prior-active-service officers join directly from the AC.

Coast Guard Model Fits and Comparisons with Cross-Sectional Data

To judge goodness of fit for the Coast Guard models, we used the parameter estimates to simulate AC and RC retention patterns by YOS under the baseline legacy retirement system and compared the simulations with the actual data to assess the extent to which the model predicts actual behavior (i.e., to assess model fit).

Figure A.3 shows the model fit for AC enlisted personnel, for entry cohorts 1994 and 2000. As we discussed earlier, we found model fits improved by allowing the value of staying in the AC for enlisted personnel at a particular point in time to vary as a function of Coast Guard enlisted end strength (end-of-year inventory). Consequently, we have model fits by cohort, reflecting the unique experience of each cohort as enlisted end strength varied over their careers, and we selected 1994 and 2000 as examples. The black line is the retention profile observed in the data, while the red line is a simulation of the retention profile under the high-three retirement system and current compensation system,[6] based on the model parameter estimates. The dotted lines are error bars. The simulations are quite close to the actual data, providing evidence that the model fits the data well. We find a close fit for the other entry cohorts (not shown). Figure A.4 shows the corresponding model fit for RC enlisted personnel with prior AC experience. Here, we find the fit is not bad, but not as good as the AC fit. That said, the Coast Guard RC force is quite small, so achieving a good fit is more challenging.

Figure A.3
Model Fit for Coast Guard Active Component Enlisted Personnel

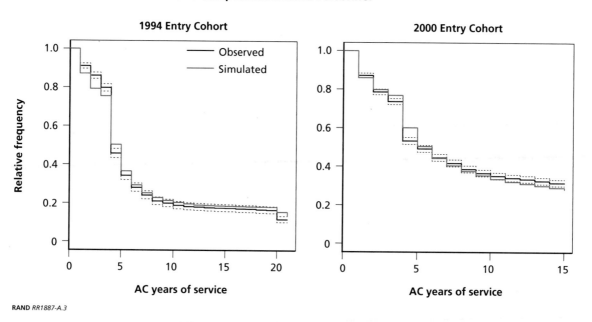

RAND RR1887-A.3

---

[6] This system has remained in place, although with some changes, over the 20-year period represented in our data, including a change in FY 2000 to allow members who entered after August 1986 to choose at 15 YOS between the high-three retirement system and the REDUX retirement system with a bonus. In the late 1990s, military pay lagged civilian pay and Congress mandated a catch-up basic pay increase for FY 2000 and higher-than-usual basic pay increases over the next six years. Higher-than-usual increases, in fact, continued through FY 2010. BAH was increased during FYs 2000 to 2005, and bonuses were heavily used from 2005 through 2008 to sustain recruiting and retention. Military retirement benefits and eligibility rules did not change but TRICARE for Life was implemented, giving military retirees continued eligibility for TRICARE after becoming eligible for Medicare.

In Figure A.5, we show the model fit for AC officers and for RC officers with prior AC experience. We found it unnecessary to allow the value of staying to vary by end strength for officers, though this was useful for enlisted. Consequently, the simulation and observed data shown in the figure are for all entry cohorts. As with the enlisted model, we find a close fit between the AC retention profile from the actual data and the simulation of the AC retention profile using the estimated model. Also as before, the RC fit is not bad, but not as good.

As mentioned, these baselines assume the current compensation system and high-three retirement system. In addition to yielding a good AC fit, we find that these baselines also replicate the retention profiles generated from cross-sectional inventory and loss data. That is, our simulated retention profiles, based on estimates that used longitudinal data on individual careers, reasonably replicate recent cumulative retention profiles constructed using cross-sectional aggregate retention and inventory data. Figures A.6 and A.7 show the comparisons of the DRM simulated baseline with the cross-sectional profiles for enlisted personnel and officers, respectively.

Figure A.4
Model Fit for Coast Guard Reserve Component Enlisted Personnel with Prior Active Component Experience

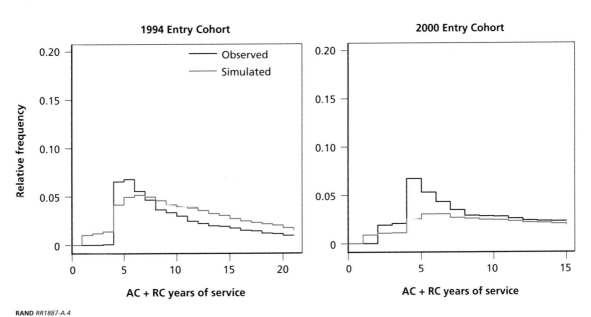

Figure A.5
Model Fit for Coast Guard Active Component Officers and for Reserve Component Officers with Prior Active Component Experience

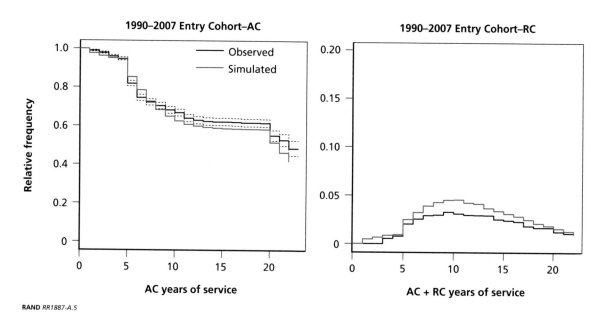

RAND RR1887-A.5

Figure A.6
Comparison of Simulated Dynamic Retention Model Baseline with 2011–2014 Cross-Sectional Retention Profiles for Coast Guard Active Component Enlisted Personnel

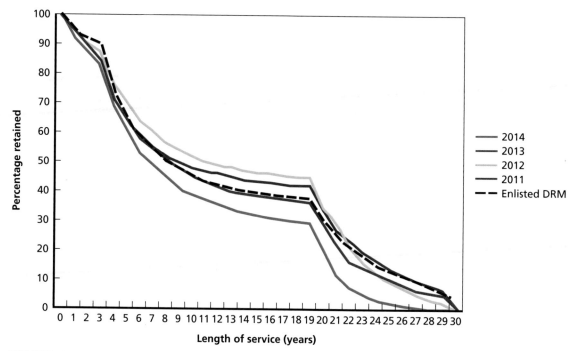

RAND RR1887-A.6

Figure A.7
Comparison of Simulated Dynamic Retention Model Baseline with Cross-Sectional Retention Profiles for Coast Guard Active Component Officers

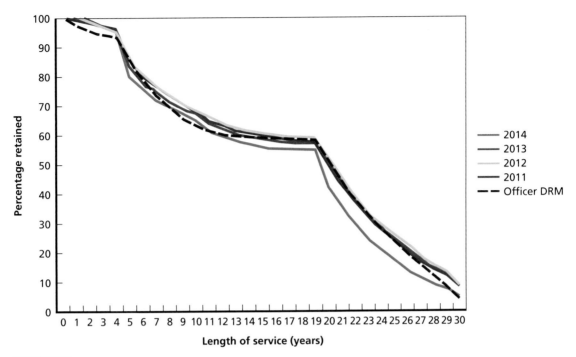

References

Asch, Beth J., James R. Hosek, Jennifer Kavanagh, and Michael G. Mattock, *Retention, Incentives, and DoD Experience Under the 40-Year Pay Table*, Santa Monica, Calif.: RAND Corporation, RR-1209-OSD, 2016. As of December 14, 2016:
http://www.rand.org/pubs/research_reports/RR1209.html

Asch, Beth J., James Hosek, and Michael G. Mattock, *A Policy Analysis of Reserve Retirement Reform*, Santa Monica, Calif.: RAND Corporation, MG-378-OSD, 2013. As of April 11, 2014:
http://www.rand.org/pubs/monographs/MG378.html

———, *Toward Meaningful Compensation Reform: Research in Support of DoD's Review of Military Compensation*, Santa Monica, Calif.: RAND Corporation, RR-501-OSD, 2014. As of November 21, 2016:
http://www.rand.org/pubs/research_reports/RR501.html

Asch, Beth J., James R. Hosek, Michael G. Mattock, and Christina Panis, *Assessing Compensation Reform: Research in Support of the 10th Quadrennial Review of Military Compensation*, Santa Monica, Calif.: RAND Corporation, MG-764-OSD, 2008. As of November 21, 2016:
http://www.rand.org/pubs/monographs/MG764.html

Asch, Beth J., Michael G. Mattock, and James R. Hosek, *A New Tool for Assessing Workforce Management Policies Over Time: Extending the Dynamic Retention Model*, Santa Monica, Calif.: RAND Corporation, RR-113-OSD, 2013. As of November 21, 2016:
http://www.rand.org/pubs/research_reports/RR113.html

———, *Reforming Military Retirement: Analysis in Support of the Military Compensation and Retirement Modernization Commission*, Santa Monica, Calif.: RAND Corporation, RR-1022-MCRMC, 2015. As of November 21, 2016:
http://www.rand.org/pubs/research_reports/RR1022.html

Christian, John, *An Overview of Past Proposals for Military Retirement Reform*, Santa Monica, Calif.: RAND Corporation, TR-376-OSD, 2006. As of April 11, 2014:
http://www.rand.org/pubs/technical_reports/TR376.html

DeNavas-Walt, Carmen, Bernadette D. Proctor, and Jessica C. Smith, *Income, Poverty, and Health Insurance Coverage in the United States: 2007*, U.S. Census Bureau, Current Population Reports, P60 235, Washington, D.C.: U.S. Government Printing Office, 2008. As of May 18, 2017:
https://www.census.gov/prod/2008pubs/p60-235.pdf

DoD—*See* U.S. Department of Defense.

Goldberg, Matthew S., "A Survey of Enlisted Retention: Models and Findings," *The Ninth Quadrennial Review of Military Compensation*, Volume III, Chapter II, Washington, D.C., 2002.

Gotz, Glenn, "Comment on 'The Dynamics of Job Separation: The Case of Federal Employees,'" *Journal of Applied Econometrics*, Vol. 5, No. 3, pp. 263–268, 1990.

Henning, Charles A., *Military Retirement Reform: A Review of Proposals and Options for Congress*, Washington, D.C.: Congressional Research Service, R42087, November 17, 2011. As of April 11, 2014:
http://www.fas.org/sgp/crs/misc/R42087.pdf

Hudson, Rex, *A Summary of Major Military Retirement Reform Proposals, 1976–2006*, Federal Research Division, Library of Congress, Washington, D.C., November 2007. As of April 11, 2014:
http://www.loc.gov/rr/frd/pdf-files/CNGR_Summary-Military-Retirement.pdf

Mattock, Michael G., James Hosek, and Beth J. Asch, *Reserve Participation and Cost Under a New Approach to Reserve Compensation*, Santa Monica, Calif.: RAND Corporation, MG-1153-OSD, 2012. As of December 10, 2014:
http://www.rand.org/pubs/monographs/MG1153.html

———, *Policies for Managing Reductions in Military End Strength*, Santa Monica, Calif.: RAND Corporation, RR-545-OSD, 2016. As of December 14, 2016:
http://www.rand.org/pubs/research_reports/RR545.html

MCRMC—*See* Military Compensation and Retirement Modernization Commission.

Military Compensation and Retirement Modernization Commission, *Report of the Military Compensation and Retirement Modernization Commission, Final Report*, Washington, D.C., January 2015. As of February 27, 2015:
http://www.mcrmc-research.us/02%20-%20Final%20Report/MCRMC-FinalReport-29JAN15-HI.pdf

Office of the Assistant Secretary of Defense, Reserve Affairs, *Official Guard and Reserve Manpower Strengths and Statistics*, Washington, D.C., 2007.

Office of the Secretary of Defense, "The U.S. Uniformed Services Blended Retirement System: Active Component," revised September 2016. As of May 18, 2017:
http://militarypay.defense.gov/Portals/3/Documents/BlendedRetirementDocuments/DoD-BRS%20AC%20FINAL%2010.1.2016.pdf?ver=2017-01-27-111503-300

Office of the Under Secretary of Defense for Personnel and Readiness, Directorate of Compensation, *Selected Military Compensation Tables: 1 April 2007, 40 Years of Service Pay Table*, Washington, D.C.: U.S. Department of Defense, 2007. As of May 5, 2011:
http://militarypay.defense.gov/Portals/3/Documents/Reports/GreenBook_APRIL_40YOS_2007_Dist.pdf

Warner, John T., *Thinking About Military Retirement*, Alexandria, Va.: Center for Naval Analyses, CRM D0013583.A1/Final, 2006.

U.S. Department of Defense, *Report of the Eleventh Quadrennial Review of Military Compensation, Main Report*, Washington, D.C., June 2012.

———, *Concepts for Modernizing Military Retirement*, Washington, D.C., March 2014.

———, *Introduction to the Blended Retirement System*, PowerPoint presentation, Washington, D.C., October 1, 2016. As of June 24, 2017:
http://militarypay.defense.gov/BlendedRetirement/